24.99

SELECTING THE RIGHT MANUFACTURING IMPROVEMENT TOOLS:

What Tool? When?

SELECTING THE RIGHT
MANUFACTURING
IMPROVEMENT TOOLS:
What Tool? When?

RON MOORE

AMSTERDAM • BOSTON • HEIDELBERG • LONDON
NEW YORK • OXFORD • PARIS • SAN DIEGO
SAN FRANCISCO • SINGAPORE • SYDNEY • TOKYO

ELSEVIER   Butterworth-Heinemann is an imprint of Elsevier

Butterworth–Heinemann is an imprint of Elsevier
30 Corporate Drive, Suite 400, Burlington, MA 01803, USA
Linacre House, Jordan Hill, Oxford OX28DP, UK

Library of Congress Cataloging-in-Publication Data
Application submitted

British Library Cataloguing-in-Publication Data
A catalogue record for this book is available from the British Library.

ISBN 13: 978-0-7506-7916-9
ISBN 10: 0-7506-7916-6

For information on all Butterworth–Heinemann publications
visit our Web site at www.books.elsevier.com

Printed in the United States of America
06 07 08 09 10 11   10 9 8 7 6 5 4 3 2 1

Contents

v

This book is dedicated to the memory of my father, and to the continuing presence of my mother, both of whom I owe so much more than I will ever be able to repay.

Acknowledgments

It is difficult to know where to begin.

Perhaps I should first thank Phil Carmichael, who prodded me into agreeing to write this book. Thanks, Phil, in spite of my continuing struggle to actually do the writing. And then thanks to Andrew and Jill Fraser whose additional prodding led me into developing much of the material used in the book.

Thanks to all those who personally took time to advise and counsel me on many of the various tools: Bill Steele for his extensive help with the RCM and PdM/CM chapters; John Schultz for his counsel and allowing me to use his material on PdM/CM and the data supporting the value of PdM/CM; Bob Williamson for his help with the TPM chapter; Ron Rath for his counsel regarding lean manufacturing, Kaizen, and Six Sigma chapters and case studies; David Burns for his counsel and allowing me to use his work on Six Sigma and engaging employees; Peter Todd for allowing me to use of his PdM flow chart; Ian Gordon for allowing me to use his work on operator ownership and maintainer/improver principles; Doc Palmer for allowing me to use his material on planning and scheduling; Bill Holmes at SIRF-Roundtables for allowing me to use his material on root cause analysis and the various tools available; Chris Eckert and Dean Gano for allowing me to use their root cause analysis material; Andrew Fraser for allowing me to use his material; David Ellison for his key insight into the material; and to Stan Grabill for his help with the Six Sigma section.

Thanks to those who helped review and critique this book. I know your time is scarce, and I am grateful to you: Vince Adorno, Wayne Barnacal, and Tim Eberle.

Thanks to all those who work day-to-day in manufacturing plants for sharing your experiences with me. Your dedication and the example you set is something I will always admire and respect. Keep up the great work that you do to make the things that we use everyday. It is a pleasure to work with all of you.

Thanks to Joel Stein, Shelley Burke, and those at Elsevier for their patience and hard work in publishing this book, and thanks to Christine Brandt at SPi for her contribution.

My last and greatest thanks are for my wife and best friend, Kathy, whose encouragement, patience, guidance, and love will always be treasured.

Preface

Manufacturing companies are deluged with various improvement tools. These are characterized as tools, methods, techniques, technologies, processes, practices, systems, and perhaps by other names that do not immediately come to mind. For purposes of this book, although not fully adequate, I will typically use the word "tools." SIRF-Roundtables, a networking organization of over 100 companies in Australia, surveyed its members and developed a listing of some 75 of these so-called tools, which are listed along with summary descriptions in Appendix C. Several additional ones that I know of are not on the list; perhaps over 100 exist. Given the availability of all the tools for improvement, why is it that most manufacturers do not seem to be getting much better, particularly in the U.S. and in European countries, which have seen their manufacturing base as a percent of GDP decline substantially over the past few decades, often to the benefit of countries like China that have lower labor costs? Are they not able to offset their higher labor costs with the improved productivity that should come with these tools? Perhaps not. There is no doubt in my mind that this inability to effectively apply and sustain these tools is due in part to corporate leadership not having what Deming called "constancy of purpose" and their strong tendency to focus on quarterly profits, something that Toyota, for example, does *not* do. Granted, you have to get the balance right between short- and long-term focus, but clearly Toyota has done a much better job at this. Witness its gains in the U.S. market against rivals such as General Motors and Ford, even while using even more U.S. manufacturing plants with their higher labor costs.

Alternatively, my experience has been that after initially applying the tools and obtaining some improvement, they far too often lapse back into old habits, losing the benefits gained. Why are they unable to use the right tool to effectively achieve and sustain the gains? Which tool is best for a given manufacturer or problem? How do we select the best tools for our situation, sustain the gains, and continue to get even better, so that we can compete in this Darwinian world of global capitalism? I hope to provide some guidance to help you answer these questions. This guidance will not be perfect, and you are encouraged to adapt it to your particular situation. There are just too many variables that can affect the outcome of a given business situation to be able to confidently pick any single tool for improvement. But, the principles outlined in this book will help you get better at it and improve your business.

All of the improvement tools I have encountered *will* work. That is, you can get substantial measurable improvement by applying them in a disciplined manner. In part, I think this success stems from the so-called Hawthorne Effect—people will work harder because they are participating in something new and you are expecting them to do better. As many of you know, in the 1920s, the Western Electric Company's Hawthorne, Illinois, manufacturing plant studied the effect of changing working conditions on productivity. For example, they speculated that if they improved the lighting in the plant, the workers would be more productive. They improved the lighting, and productivity did improve. Then, to test their theory, they dimmed the lighting, thinking productivity would decline. But, the opposite happened—they dimmed the lighting and productivity improved even more. To summarize their conclusion in my words, if you give people a little attention, and they know you are expecting them to do better, they will. I think an old principle applies here—no matter what tool you use, if people know you are looking for them to do better, they will work hard with that tool to meet your expectations. This effect is similar to the so-called Pygmalion effect described by J. Sterling Livingston, that is, high expectations lead to the development of a "super staff," whereas low expectations increase the risk of failure. People in a subordinate position will work hard to meet your expectations, so long as the expectations are reasonable and they have the time, tools, and training to do the work required.

Nonetheless, the question still remains as to which tool is best in a given situation. Should we adopt a lean manufacturing model patterned after the Toyota Production System? Toyota is among the best manufacturers in the world. Or, is using Six Sigma a better approach? After all, Jack Welch was famously successful in applying Six Sigma at General Electric. If we use Six Sigma, the logic is, we will be more successful. Moreover, when should we use some of the other common tools, such as supply chain management, Kaizen, total productive maintenance (TPM), reliability-centered maintenance (RCM), root cause analysis (RCA), or any other of a number of tools for improvement. Are the tools compatible? Or, are they counter to one another? Which ones *are* incompatible? The question we must ask is which tool provides the greatest value in light of our current business situation and our strategy and goals for the future.

Coming across many of these tools in my day-to-day work with manufacturing companies, along with the prodding from two friends, led me to decide to learn more about them and to subsequently share that in this book. I must emphasize that *I am not an expert in these tools.* But I have had considerable exposure to all of them and will share what I have learned, both in my studies and in the field while working with manufacturers who have used these tools. While my work will typically summarize or highlight the key practices and processes of the major improvement tools, I will also provide lots of references that delve more deeply into the tools. *What I hope to do is to provide enough information so that you can relate your circumstances and issues to the tools and make better decisions as to what is relevant and beneficial to you.*

As noted, my experience has been that all the tools work, when consistently applied over the long term. What is more important in my opinion is the leadership aligning the organization to a common strategy and set of goals, creating a culture of teamwork and continuous improvement, and then selectively applying the appropriate tools. But, some tools may not be as good as others for a given situation. For example, rigidly applying the lean manufacturing principles (the Toyota Production System) and/or TPM to a process manufacturing plant, such as an oil refinery or smelter, is probably not the best approach. The lean manufacturing principles are excellent and encouraged for *all* manufacturers. However, in a process plant, they are more difficult to literally apply since they were developed around automotive, or discrete, manufacturing. Process manufacturers are typically inherently leaner and have more "level flow" in production, a key element in applying lean principles. They tend to run more steadily and do not keep much product/inventory in pipes or tanks; the cost and risk of starting up and shutting down an oil refinery or steel foundry, for example, give one cause to reflect before applying tools like takt time and Kanban, which facilitate making only to demand. These operating conditions in a process plant often require considerable adaptation when applying lean principles. Alternatively, TPM is likely to be a better tool for a discrete plant than RCM. Similarly, launching a Six Sigma program when your basic operating and maintenance practices are very poorly done, leading to considerable variability and instability, is probably not a good decision. It is essential to get the basics right first. Doing so will substantially reduce the variability in your processes. You can then apply the Six Sigma program with much greater efficiency and effectiveness. In any event, no matter what tool is appropriate, it is always a good idea to get the basics right first. Doing so will substantially eliminate the need for any particular tool and make any tool you apply more effective.

My experience has also been that discrete plants, such as automotive plants, tend to benefit *more* from using lean principles and from Kaizen, 5S, and TPM, and tend to benefit *less* from RCM, PdM, and maintenance planning and scheduling, although RCM can be used to finetune the maintenance practices called for in TPM. Discrete plants will benefit from all of these tools, only less so from the latter. Discrete plants also typically have an operator to maintainer ratio of ~10:1, more or less, making the role of the operator relatively more important to manufacturing excellence than in process plants. Process plants on the other hand tend to be benefit *more* from RCM, PdM, and maintenance planning and scheduling and *less* from lean principles and tools, and they typically have an operator to maintainer ratio of 3:1, more or less. This makes operator care and attention to detail relatively less important in a process plant. Make no mistake: it is very important in both types of plants, so we are dealing with relative influence, not absolute.

Discrete plants also tend to have much lower maintenance budgets than process plants, often 20% of that for a process plant with a similar number of employees. For example, a typical automotive plant with 400–500 people might have a maintenance budget of $6M, whereas a process plant with the same number of employees might have a budget of $30M. Please understand that all of the tools and strategies can be applied at some level for all of the plants, but getting the right emphasis on the right tools is key to achieving the greatest benefit. Further, the strategic value of gross profit improvement from measuring and managing overall equipment effectiveness (OEE) tends to be much greater in discrete plants than in process plants. For example, at one automotive plant, the value of improved reliability on gross profit improvement with 33% reduced maintenance costs was about $2M, but the value of improved OEE/production on gross profit, presuming market demand was available, was ~$50M. Even if demand was constant, the value of reduced costs through higher productivity was ~$10M. On the other hand, at one process plant, the value of reduced maintenance costs through improved reliability and productivity was ~$10M, while the value of improved gross profit was ~$10–$20M, depending on market demand and operational productivity improvement. In any event, all of the principles and tools apply to all plants, but the degree to which we apply them will vary considerably depending on the type of plant and the business model being used for its operation.

The content of this book is written from the perspective of an observer or erstwhile practitioner. My personal experience has been that most employees want to do a good job. A few, typically <5%, do not. We need to sort them out and manage them, but not to the detriment or burden of the >95% who do want to do a good job. More importantly, we need to develop systems that bring out the inherent good in people, by supporting and challenging them and helping them to be tenacious and disciplined in getting the basics right: (1) operating with care, within the inherent capability of the system, assuring consistency across all shifts, and excellence in process conformance; (2) maintaining with precision and craftsmanship and applying discipline in work management; (3) working as a team with a proactive view to eliminate defects at their source, while keeping our business objectives in mind; and (4) measuring our success. The best companies do all the basics really well, enabling more time for both shop floor and business level innovation, forward thinking, and ultimately continued success. In doing this, we must continuously develop managers that constantly, but supportively, challenge the status quo, that demand excellence in our practices, that do not accept mediocrity, and that foster an environment of continuous improvement.

This book describes several of the more common and popular methods, including case studies, being used by manufacturing companies to improve their business. Even before we cover these tools and strategies, we cover several fundamental issues—if present, almost any tool will be effective, but if absent, almost no tool will work, i.e., the need for leadership and several models thereto for creating an environment that brings out the best in people, the need for aligning the organization and how that might be done, the need for innovation *throughout* the organization, including what I call "big innovation"—R&D and new product/process development, but also "little innovation"—using the proper tools in a disciplined way to lower our costs and improve quality so that we have higher margins to pay for the big innovation. We will also cover the use of cross-functional teams in applying these methods, and a process for managing change, as well as a model for measuring performance and demonstrating that operational and business goals have been achieved. Finally, we will review a process that I call business level failure modes and effects analysis (FMEA) for identifying the right projects for improvement, and for selecting the right tools for that project.

More specifically, we will review the following:

- Lean manufacturing
- Kaizen, including 5S, Kanban, quick changeover, and standardization
- Total productive maintenance
- Six Sigma
- Supply chain management
- Reliability-centered maintenance
- Predictive maintenance (or condition monitoring)
- Root cause analysis
- Additional information will be provided in the appendices regarding:
- Planning and scheduling
- Performance measurement
- Other tools

There are numerous tools that are quite good, but will only get a brief mention in the appendices, e.g., The Manufacturing Game®, Kepner

Tregoe®, and others. It is not that these are not useful, for I believe they are, especially The Manufacturing Game®, which is an excellent, experiential change agent for engaging the shop floor. Rather, time and other limitations did not allow me to cover these in any detail. And, I will likely miss a few tools. Those who feel slighted because of this, please accept my apologies. Perhaps we can cover those in a future edition of the book.

Someone once admonished me that the best way to teach something is to tell a story. So, in doing all this, we will continue to use a fictitious company called Beta International as a "vehicle" to communicate the various messages and case studies. As some of you may recall, we used Beta in my previous book titled *Making Common Sense Common Practice: Models for Manufacturing Excellence* to outline a process for improving manufacturing performance by getting the basics right. In this book, Bob Neurath, Beta's CEO, will continue to play a visible role in providing leadership, managing change, and aligning the organization to a common strategy and set of goals. I will take certain figures and text from the previous book and story about Beta to illustrate and reinforce several key points, but often from a different perspective. I will repeat some of those messages within this book to reinforce certain points, in keeping with an old Russian proverb—repetition is the mother of learning. I will also use Bob's character to offer certain opinions that while perhaps not yet a part of mainstream thinking, I believe can benefit most, if not all, companies. We will continue to use Beta in this book to illustrate real case studies that reflect the experience of various manufacturing plants, but the descriptions have been modified to mask the identity of the plants. In telling Beta's story, at times I will use what Robert Fulghum's called his story teller's license. That is, I may, like Fulghum, "...(rearrange) my experience to improve a story so long as it serves the standard of truth in the best sense. The truth of poetry and parable do not compete with the truth of science or the courtroom. I trust the reader to know the difference." As such, although based on real experience, these case studies are not intended to describe any specific company's actual performance. Therefore, any correlation, real or imagined, between Beta and any other company is coincidental. Beta International will continue to be used to illustrate the composite experience of many different companies.

Someone once said "I write to teach myself." Likewise, I have learned a great deal in writing this book, and so much from so many of you who offered your guidance. However, I certainly know less today than I will in the future as you and your colleagues review and sometimes challenge my work offering further enlightenment. Thank you for buying this book. I hope it will help you better apply many of the more common improvement tools more appropriately and with a greater degree of success.

Ron Moore
Knoxville, TN

Introduction

1

Productivity isn't everything, but in the long run it is almost everything.

Paul Krugman, Economist

The Scene

Beta International, a large manufacturing conglomerate, has made substantial progress in its operations and business performance. A few years ago, Bob Neurath, CEO, set forth a 10-point strategic plan for process improvement that focused on getting back to basics.[1] After all, he reasoned, if we aren't tenacious about getting the basics right, how can we expect to get anything else right? Driven from the top, this focus has yielded substantial gains for the company. Asset performance has improved, unit costs are decreasing, safety performance continues to improve, and market share is increasing in many markets, holding steady in others, and only declining in a few markets that are on strategic watch. Profits have moderately increased in most business units and holding steady in almost all others.

But, performance is simply not good enough. Return on capital is only about average in most business units. Intense pricing pressures from foreign competition, particularly Asian manufacturers, are continuing to drive down prices and hold down margins. If downward pricing pressures are stemming from Asia, Bob muses, then there is no doubt that companies like Wal-Mart, with their guaranteed lowest-price strategy, will only further intensify those pressures. The low margins resulting from these pressures limit Beta's ability to invest in research and development (R&D) for new product and process development, new capital projects that flow from this R&D, as well as strategic acquisitions better aligned to the company's long-term strategy and business objectives. As a result, future market development and the growth that goes with it are likely to be hampered. In summary, the good news is that Beta has returned from the brink of disaster to become at least a mediocre company, maybe even above average in some business units. The bad news is that Beta is still not much better than a mediocre company. Perhaps more importantly, many seem to be satisfied, even pleased, with this. *Bob Neurath is not.* Having taken a step back from the abyss it faced a few years ago, Bob is now ready to move Beta forward more aggressively, creating a much more profitable future for the business. Trying to be all things in all markets only dilutes management focus, thus leading to mediocrity. Bob's expectation, like many corporate leaders today, is that Beta should be first or second in all of its markets or have a clear, measurable path for achieving that position. The company must now move from survival mode to growth mode.

Beyond the "back to basics" focus in all practices for all operations, several of Beta's divisions have tried various improvement tools with varying degrees of success. One division was convinced that Six Sigma would provide substantial gains and achieved substantial improvement in some of its operations, but not others. Another division considered total productive maintenance (TPM) more appropriate and likewise achieved substantial gains in some operations, but not others. Another adopted the Toyota Production System—what has come to be known as Lean Manufacturing—or so they thought. In that division, Bob's characterization is not that they are lean and mean, but rather anorexic and angry. Although improvements were achieved, overall results were less than desired, and on further review, it appears many of the initial results have been lost and they may have left out key strategic and philosophical elements of the system.

Beta's experience has been similar to that of other companies. Various tools or methods are touted as a means for a major step forward but never seem to quite deliver the expected results. For example, Bob Williamson, an industry expert in TPM, estimates that some 60% of TPM programs fail after three years of effort; Jeffrey Liker, an industry expert in Lean Manufacturing, estimates that less than 1% of U.S. companies are truly effectively applying lean principles.

Neil Bloom, an industry expert in reliability-centered maintenance (RCM) estimates that over 60% of all RCM programs initiated are never successfully implemented, with many of the rest only done in a superficial way. RCM is an excellent tool for identifying and minimizing functional failures, is also on occasion called the "resource consuming monster." Given the limited resources in most companies through downsizing, RCM does not typically get the support it needs to do the analysis or the results end up being stored in binders on a shelf. Root cause analysis has various approaches, including the "5 Whys" supported by Toyota and progressively more comprehensive approaches, but it can also be very resource-intensive, limiting management's ability to successfully apply it.

The Response

Bob's response to this lack of consistency and "constancy of purpose" (Deming) in applying these tools, beyond a certain amount of chagrin, was to form a senior-level steering team to review each tool and make a judgment about its efficacy and efficiency. Some questions being asked were: (1) under what circumstances do we apply each strategy or tool? (2) what are each tool's advantages and disadvantages? (3) are they compatible and/or supportive? (4) do some require other related tools to be used in conjunction for maximum results? and (5) when are they incompatible? These fundamental questions need answers in order to develop a strategy for their application that could align the organization toward a common set of goals within all of Beta's various operating units.

Bob also demanded an annual market review to ensure proper alignment of the manufacturing plants' strategy to the marketing and product mix strategy (and vice versa). The review needed to identify the major steps needed for assuring growth in all markets. He agreed with Michael Treacy's view in *Double Digit Growth*[2] that growing market share required Beta to do most, if not all, of the following seemingly straightforward activities:

1. Keep the customers that you already have by creating incentives for them to stay or obstacles to their departure.
2. Take business from your competition by making your products more attractive.
3. Go to where the growth is by anticipating those growth areas.

4. Enter adjacent markets by adapting your products to those markets.

5. Invest in developing new product lines, markets, and customers.

These principles need to be applied at Beta, along with the models for rationalizing and optimizing products, customers, and markets outlined by Christopher[3] and Moore[1] to better manage those products, customers, and markets and, particularly, to align these to the manufacturing capability. Both recommend a routine analysis of customers and products and their contribution to the business. The model used by Christopher suggests dividing customers and products into A, B, and C categories. For example, the A list is the 20% of customers who provide 80% of profits and/or volume or the 20% of products that provide 80% of profits and/or volume. The A list must be sustained and built over time. The B and C lists require either development or elimination. Both Christopher and Moore offer models for optimizing customers and products.

Bob will continue to use the model shown in Figure 1-1 to help align the marketing and manufacturing strategies.[1]

The marketing department will manage product and customer mix and complexity within the product range. They will work with manufacturing to understand the implications of their decisions on the manufacturing function, particularly as it relates to quality, cost, and

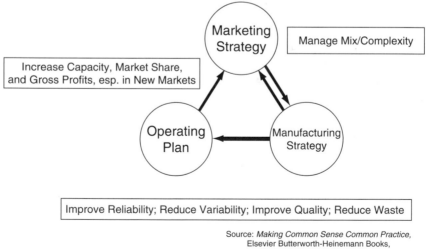

Source: *Making Common Sense Common Practice,*
Elsevier Butterworth-Heinemann Books,
Boston, MA, & London, England

Figure 1-1. *Aligning the Marketing and Manufacturing Strategies.*

delivery. This communication will be a two-way street with business decisions based on what is best for the business overall and not on a single given function. The plants work to improve quality, cost, and delivery through improved reliability and stability as well as reduced variability and waste. This should increase capacity and gross profits and allow the pursuit of additional market share without having to make additional capital investment.

Bob also agrees with Bossidy and Charan's contention that: (1) the world is awash in capital creating overinvestment and excess capacity; (2) China and India are attracting much of that capital with their large markets and cheap labor; and (3) when combined with the Wal-Mart/Home Depot business model for driving down prices, costs must follow downward.[4] Most importantly, Beta must compete in that environment. Beta must become more productive, reduce its waste and system/supply chain cycle times, and use its inherent advantages. It would also require Beta to take stronger positions in the Chinese and Indian markets, both in terms of products and customers, as well as potential manufacturing capability to address those markets and perhaps others worldwide. Clearly, this will all be challenging.

Bossidy and Charan's model for doing this is shown in Figure 1-2. They suggest that we must fully understand our business's external realities: Are we in a growing or declining market? What is happening to prices in those markets? What is happening to our customer base? Are we in the middle of a major structural shift? They also suggest that

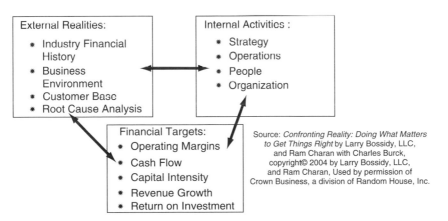

Figure 1-2. *The Business Model.*

we understand our internal activities relative to our external realities. Is our strategy aligned with market realities? Are our operations efficient enough to address those? Are we as executives engaging our people in supporting our improvement? Is our organizational structure aligned to meeting our objectives? And, are we meeting our financial targets in light of all this, routinely iterating and improving our performance relative to the external realities and internal activities?

Bossidy and Charan offer a number of suggestions and case studies in addressing these issues, something that Beta must adopt to be more effective in managing its business.

The Tools

An initial review of the tools and strategies being used in various Beta operating units revealed that the following were applied to some significant degree, sometimes in conjunction, but at other times independently from one another. That is, different plants or functions were applying one tool or the other without coordinating their application. In any event, the most common tools and strategies being applied were found to be:

- Lean Manufacturing
- Kaizen
- Total Productive Maintenance
- Six Sigma
- Supply Chain Management
- Reliability-Centered Maintenance
- Predictive Maintenance
- Root Cause Analysis

Some, such as Lean Manufacturing and Kaizen, included the use of tools of Total Quality Management/Assurance (TQM/A), 5S, quick changeover, and other techniques. Others, like Six Sigma, included the use of statistical process control techniques and design of experiment (DOE). A summary of the steering team's initial findings regarding each of these is provided next.

Lean Manufacturing

At its heart, *Lean Manufacturing focuses on the elimination of waste in all forms*. Categories of waste include excess inventory, wait and delay times, off-spec product, and so on. Several of Beta's operating units have adopted or, more accurately, adapted the Toyota Production System, or what has come to be called Lean Manufacturing, as a model for improving their manufacturing performance. This has been done with mixed results: some plants have gotten better, others have stayed about the same, and some have actually gotten worse. It also appears that the process manufacturing plants in Beta's operations have had more difficulty applying lean principles. They have less ability to meet demand or do "pull" production, and therefore, they tend to "push" production. This tendency to "push" generally relates to a lower ability to operate efficiently, if at all, at production rates that are much below design or to the difficulty and risk in shutting down and starting up the plant over short periods of time. But this condition also provides for level flow, a key Lean Manufacturing condition. For example, at an oil refinery, you cannot simply stop producing from one day or one shift to the next, so the times for applying "pull" to your production and level your flow tend to stretch into weeks or even months. The process plants have also tended to be inherently leaner than batch and discrete plants; they don't or cannot carry a lot of intermediate stocks to compensate for disruptions. And, as pointed out earlier, they have a fairly level production flow, a key advantage in applying lean principles. The process manufacturers generally also had better reliability, better operating and maintenance practices, and better plant management support for manufacturing excellence.[1] After all, in many process plants, if one step in the process fails, the entire plant often goes down. So, there is a lot of pressure to keep the plant running and implement the correct practices. Whereas in a discrete plant, if one machine fails, then it often happens that people believe that loss can be somehow "made up." Unfortunately, once time is lost, it is lost forever.

The concept of Lean Manufacturing first came to be more widely known with the book, *The Machine that Changed the World: The Story of Lean Production*.[5] This book is an excellent review of the history of the development of Lean Manufacturing, primarily at Toyota, but is scant on the details of the methods for achieving it. More recently, two books—*Running Today's Factory: A Proven Strategy for Lean*

Manufacturing[6] by Charles Standard and Dale Davis and *The Toyota Way*[7] by Jeffrey Liker—have provided a much clearer description of Lean Manufacturing principles and, more importantly, how to apply them. We will discuss these principles more in later chapters.

Unfortunately, the apparent behavior of many of Beta's managers, as evidenced by the performance at their manufacturing plants, suggests that the books by Liker and Standard and Davis have not been widely read or practiced. For example, and over-simplifying a bit, after one management team's trip to a plant reputed to be excellent at Lean Manufacturing, the team returned to their plant thinking that the lean plant did not have much inventory or people. So, the following reasoning surfaced: If we reduce our people and inventory, we will be lean too. Of course, this ignored the "hidden" processes and systems that were in place at the lean plant that allowed them to have less inventory and fewer people to accomplish the same result. The consequence of this thinking was that as costs went down (in the short term) so did delivery performance and quality, to the chagrin of many customers.

One of the major issues that must be addressed is the understanding (or lack thereof) that Lean Manufacturing is *not* focused on head-count reduction. Lower head count per unit of output is a consequence of applying lean principles, and yet it seems that many managers even at very high levels still believe that reducing head count will result in a lean company. There is a large body of evidence, some of which is discussed in later text, which states the contrary. It is essential to fully understand lean principles. An analogy which may help is to recognize that "lean" and "fit" are two very different concepts. Olympic athletes are "fit," which would typically give them a "lean" appearance. Anorexic people are "lean," but hardly "fit," and we certainly do not want an anorexic company. Another analogy is Gary Hamel's; he coined the phrase "corporate liposuction" to describe how some companies use cost cutting to become "lean" in the short term, only to suffer significant deterioration in performance over the long term.[8] Finally, Deming stated, "Your system is perfectly designed to give you the results that you get," a very self-evident statement, and yet very profound. A corollary to Deming's statement would be: If you reduce the resources available to your system without changing its basic design, then system performance will decline. Before we take a brief look at the other tools being employed by Beta, let us have a look at the concept of cost cutting and its efficacy.

Cost Cutting—Does it Work?

Data from studies of companies that have engaged in cost cutting and particularly head-count reduction supports the corollary to Deming's quote. For example, in one study the following data were reported for several hundred companies that engaged in cost cutting through layoffs or downsizing over a period of about five years[9]:

- Only 50% showed profit improvement.
- Only 34% showed productivity improvement.
- Some 86% experienced a serious decline in morale.

Similar research conducted by the U.S. Conference Board[10] on companies using a downsizing strategy reported that:

- 30% actually experienced increased costs.
- 22% terminated the wrong people.
- 80% reported a collapse in morale.
- 67% showed no immediate productivity rise.
- 50% showed no short-term increase in profits.

They also pointed to another concern: the loss (or delay) of technology that results from the inevitable cutbacks in R&D staff during tough times. They found that six months after a layoff, there was an increase in share price relative to market indices, but that after three years, share prices had declined relative to market indices.

As previously noted, Hamel refers to cost cutting as "corporate liposuction" (all kinds of unpleasant images also come to mind with this well-chosen metaphor). He defines this as a condition in which earnings growth is more than five times the sales growth, generally achieved through cost cutting. In a review of 50 Fortune 500 companies engaged in corporate liposuction, 43 suffered a significant downturn in earnings after three years. Hamel points out that growing profits through cost cutting is much less likely to be sustainable and must be balanced with sales growth through efforts such as innovation, new product development, and process improvement.[8]

And finally, Morris, Cascio, and Young reported that in a study of 3,628 companies reviewed over a 15-year period.[11]

- "Employment downsizers don't improve financial performance."

- "...those with the largest layoffs exhibited the largest decreases in ROA [return on assets]."

- "One striking aspect of downsizing is that the impact on profitability is negligible relative to the magnitude of the layoffs."

- "...not only did they fail to increase their return on assets, but they experienced a continued decline on their return."

This is not an all-inclusive data set since others have reported similar findings. In summary, cost cutting is highly problematic in reducing costs and/or improving overall performance.

The aforementioned text does not imply that cost cutting never applies. From the previous data, cost cutting is not likely to lead to prosperity. However, it may work some of the time, and it has a great deal of uncertainty associated with it. Not all of the companies studied suffered as a result of their cost cutting. So, when does it work? Perhaps under the following circumstances[1]:

- If the company is near bankruptcy, and you have no choice but to "stop the bleeding" to survive.

- If the company is a bloated bureaucracy, and you must strategically address your cost structure before lean, mean competitors begin taking your market share.

- If you are faced with intransigence in employees, unions, and so on and/or need to get people's attention about improving productivity to assure a sustainable competitive position.

- In specifically targeted situations (e.g., obvious waste or when two companies with redundant functions or roles merge and consolidation is needed). These benefits may be slight, however, and may be overwhelmed by the difficulty of effectively integrating the cultures of two organizations and aligning the new organization to a common strategy and set of goals.

- In a severe market disruption or downturn (e.g., 10%–20%) reducing industry volume, or when dramatically new technology that eliminates the need for a given product.

Some of these may overlap, or they may all be present in a given business. So we need to ask whether these may apply and to what degree. Further, not surprisingly, Morris, Cascio, and Young also reported that downsizing is most likely to work when it includes a major restructuring of physical assets. Cost cutting has a place, but the risk of cost cutting on the overall system performance must be considered. Costs and business results are a consequence of our business system design, that is, our processes.

Why Do Executives Persist in Cost Cutting?

The data suggest that *at best* cost cutting provides a 50% chance of improving company performance (*at worst,* 14%). Why then, is cost cutting viewed as one of the primary means of improving financial performance even if it has a low probability for success? It may be necessary under specific circumstances, but often it is not. So, why do senior managers often persist in enforcing cost cutting? Possible reasons include:

- Executives don't view layoffs as a possible admission of personal failure. (Then, why were all of those people on the payroll in the first place?)
- It demonstrates a bias for action.
- It is easy and simple, and it does in fact sometimes work.
- It does not require a lot of leadership skill; anyone can cut costs by 5% and expect employees to get the job done for 5% less.
- It can pressure people into being creative about doing their jobs more efficiently and eliminate waste, and that can sometimes work.

- "Everybody does it"; it is a generally accepted practice that is often warmly received on Wall Street with its focus on quarterly results.

- Executives are overconfident in believing they can beat the odds.

- It at least represents competitive parity, since most companies seem to do it. (But don't we want a competitive advantage?)

- When combined with a restructuring of physical assets and business strategy, it is more likely to work.

- Executives have never personally experienced a layoff. Perhaps if they had, they might have greater empathy for the pain they are causing.

- Perhaps most importantly, they often don't fully appreciate that costs are a consequence of your business system design. If you don't change your fundamental business system design, which can take years, but you remove resources from the system, performance will most likely decline.

Unfortunately, however, the data suggest strongly that, as in Las Vegas, the odds are inherently against these senior managers and built into the systems already established in the company, which, in turn, suggests that changing the business system design is a better strategy. Initial impressions of other tools Beta has used to "change their business system design," again with mixed results, are provided below.

Kaizen

Kaizen is a Japanese word literally meaning change for the better, or, more appropriately, continuous improvement.[12] As with the other tools, the results of using Kaizen in the various operations were mixed. Several of Beta's plants have employed Kaizen events and have seen dramatic improvement shortly after the event. However, these improvements appear to have been short lived, since within a few months of the event, performance was near what it had been before. If Kaizen means continuous improvement, why hasn't Beta sustained these improvements and gotten even better using Kaizen? Beta must not be applying Kaizen properly.

Total Productive Maintenance

Several of Beta's plants have used TPM,[13] again with mixed results. Some encountered union resistance to TPM, who referred to it as "operators doing maintenance work" or expressed negative feelings toward yet another Japanese technique. Still others, particularly the process plants, had difficulty using TPM because it seemed to be more suited for plants where the operator had autonomy over a specific, localized process, like a machining operation versus having autonomy over an area, such as the oxidation process in a chemical plant that had numerous pieces of equipment spread over a large geographic area.

Other plants did fairly well in using TPM. These plants seemed to be the ones where maintenance was defined as maintaining the plant or process function, not just on fixing things. These plants referred to TPM as "total productive manufacturing" about maintaining the plant and/or equipment function, not just repairs or preventive maintenance (PM). These plants also seemed to view it as a support tool for Lean Manufacturing, not just another stand-alone tool. "When should we use TPM?" Bob asked.

Six Sigma

Several of Beta's operating units have adopted the Six Sigma[14] tool for improvement. This tool uses concepts similar to Deming's plan, do, check, act, or more specifically DMAIC—define, measure, analyze, improve, control—on core processes and key customers. One author has described Six Sigma as "SPC on steroids." While it is literally a statistical term that characterizes your quality having <3.4 defects per million for a given product or process specification, it has become a methodology for reducing the variability of processes such that the result is greater quality and consistency. It also stresses simultaneously achieving seemingly contrary objectives (i.e., being stable and innovative, seeing the big picture and the details, and being creative and rational). At one of Beta's plants, they have been very successful applying it, with cost reductions of over 15%, but at another, the results have been dismal—over $1M spent with little or nothing to show for it. The question of course is "Why the difference?" Bob Neurath is determined to find out.

Supply Chain Management

Most of Beta's plants were using some sort of supply chain management tool,[3,15] yet again with mixed results. In some cases, using supply chain principles was viewed as putting pressure on suppliers to provide lower costs (primarily through lower prices), better quality, and better delivery performance. In some cases, this included vendor stocking. These plants seemed to be getting good results, that is, they would report big savings in supply and raw material costs, but somehow those savings seemed to evaporate when the financial statements came together. For example, the vice president of purchasing in one of the divisions reported saving $200M in lower supply costs, but the gross profits and operating costs for that division hardly budged. The vice president provided very detailed lists of prior costs and volumes for certain supplies and current costs for the same material, demonstrating the savings. Apparently these savings were being offset by increased operating and maintenance costs, since the overall profitability had not changed. They were not getting any worse, but they were not getting any better either, which really meant they were getting worse.

The divisions that seemed to be doing better in using supply chain principles appeared to view the supply chain more holistically and literally; it is a supplier, a manufacturer, and a customer, working as a team in a supply chain to optimize "chain" performance. They reviewed the entire "chain" and used process mapping to predict the chain's performance in areas like business system cycle times, inventories, distribution requirements, costs, risk/delays, and so on, working to optimize the chain, where savings were shared among chain members. Others also took a more holistic view, looking to consolidate suppliers and work in greater partnership with them, and in doing creating a more detailed review of customer requirements and rationalizing certain customers and/or products. Bob Neurath wanted to know more about so-called "best practice," particularly the best sustainable results in applying supply chain principles.

Reliability-Centered Maintenance

Still other Beta plants were using RCM as a principal means of improvement, once again with mixed results.[16] After some time, a few plants just considered it too burdensome to have 5–10 guys sitting in a room analyzing and endlessly debating equipment, system functions, and failure modes for *all possible* functions and failure modes (managers sometimes failed to send the requested people for the analysis); then having to prioritize all the failure modes using a criticality analysis; and then developing tasks for addressing the most critical failure modes. The real "kicker," however, was often *not* doing what resulted from the analysis because of limited resources, existing operating rules, union objections, state regulations, and so on. What is the point, they seemed to ask, in spending all this time and money doing the analysis, when we aren't going to do what it tells us we should do? Many times the analysis ended up as big notebooks on a shelf with little further action taken.

But, other plants seemed to being doing well using RCM. These seemed to take a more focused approach, often limiting their analysis to known failures in critical equipment that were causing the biggest problems to the plant. Since the primary objective of RCM is to *preserve system function*, they reasoned that the greatest benefit would come from analyzing those failures where the greatest loss of function had occurred and then being very disciplined about eliminating or managing the failure mode. Bob wondered, when should we be using RCM, and how do we make sure we use the results to get better business results? Should we use RCM preferentially to TPM, or vice versa? When?

Predictive Maintenance

In trying to understand predictive maintenance, Bob asked, "does predictive maintenance let you predict the time when a machine is going to fail so you can plan around that failure?" Unfortunately, the answer is no. Predictive maintenance will not allow you to predict the timing of a failure with any accuracy. Then why is it called predictive maintenance? As with so many things, the answer seems to have been lost in history and happenstance. Predictive maintenance can more accurately be characterized as a sophisticated inspection that allows you to understand the condition of your equipment, whether or not a defect is present, and whether the defect is severe enough to require timely action. And, in general, we can also apply a simple rule—combining severity of condition and consequence of failure will allow us to judge our priority for action. The greater the severity and the greater the consequence, the higher the priority will be. Which of the predictive maintenance tools is best? As you might expect, it depends. What are the most common and consequential failure modes associated with the machine? What is their notional frequency? Answers to these questions will allow us to better select the appropriate monitoring tool and frequency of data collection and trending.

How has Beta fared in applying predictive technology? Here the answer is pretty clear—very poorly. A good benchmark is that about 50% of all corrective maintenance should be done as a result of some form of condition monitoring or inspection, including operator tasks, the result of which is used to plan and schedule the work so as to minimize the consequence (maintenance cost and production downtime losses) of the pending failure.[1] At Beta, typically 40%–50% of all maintenance is done reactively, and this is better than it had been. That is, it is typical that less than a week's notice is available before the maintenance must be done, and nearly 10% of the time it is an emergency—the maintenance must be done *now*. Clearly, Beta has failed to apply these tools effectively. And, even at those plants that do a somewhat better job with predictive maintenance or condition monitoring, their planning and scheduling capability is pretty poor, so the equipment runs to failure anyway. There just seems to be a lot of opportunity here.

Root Cause Analysis

Beta's use of root cause analysis tools has been, to be kind, spotty. There does not seem to be a mindset in the company for getting to the root cause of problems. Most seem to be doing well to simply put counter measures in place to get through a given day and have little time to solve problems to root cause. Granted, some plants have had a few people trained in one method or another. A few plants even have a person trained as a facilitator in root cause analysis. But there does not appear to be a consistent process for when to do such an analysis, what method to use, or how to track any results. Bob reasoned that having a systematic process for: (1) the criteria for when to do root cause analysis, (2) actually doing the analysis, and (3) assuring the results were implemented and sustained, was critical to the future success of the business. He had heard that Toyota tended to shun the more complex root cause analysis tools, favoring a simple

5 Whys approach. In fact, it seemed too simple to him, but he was open to being convinced.

Other Tools

Thus far, this chapter has discussed the tools and strategies most commonly used in Beta's operations. There was a smattering of other tools in use at various plants. These included The Manufacturing Game®, Kepner Tregoe®, and several others. These are briefly discussed in Appendix C.

One tool that Beta will continue to use, but will not be discussed in this book in any detail beyond this section, is benchmarking, which according to Dr. Jack Grayson,[17] involves "seeking out another organization that does a process better than yours and learning from them, adapting and improving your own process...." Benchmarking, then, involves emulating the processes and practices of the other companies to improve yours so that you can improve performance as measured against benchmarks. Benchmarking is a continuing process, requiring constant attention to the latest improvement opportunities and the achievements of others. Benchmarks, on the other hand, have come to be recognized as those specific measures that reflect a best in class standard. Best practices, as the name implies, are those practices that are determined to be the best for a given process, environment, circumstance, and so on, and allow a company to achieve a benchmark level of performance for a given performance measure.

Be cautious about using benchmarks. Benchmarks (the numbers) don't tell you anything about the learning (e.g., the systems, processes, and practices that went into achieving a particular level of performance). Achieving a benchmark level of performance requires considerable learning—your processes and practices have to be exceptionally good. Simply obtaining a comparative set of numbers from other companies and then making arbitrary decisions about achieving those may actually be detrimental. For example, arbitrarily reducing spare parts to achieve a "world class" level of spares in the store room could dramatically affect plant performance in a negative way to the detriment of the overall business. If the processes are right, then the measures will improve. Four additional cautions about using benchmark data are: (1) there is considerable scatter in the data used in benchmarking, so it may be difficult to pick the "right one" for you; (2) benchmark data is constantly changing as plants improve their processes; (3) no single benchmark should be used to make decisions—the performance measures tend to be inter-dependent and all the relevant measures must be considered in light of the company's overall business goals; and (4) variables related to product mix, location, process design, and so on, will affect benchmarks. So, considerable judgment must be used in applying benchmarking principles. A table of nominal benchmark data for manufacturing operations is provided by Moore[1], and a process for performing benchmarking studies is provided in *The Benchmarking Management Guide*.[18]

Summary

It seemed to the steering team that all the tools would work, if con-sistently and properly applied, and that most all had certain elements in common. For example, typical goals were: (1) to minimize costs by reducing waste, in all its forms—excess inventory, poor quality, downtime, and so on; (2) to assure consistency and/or reduce the variability in plant processes; (3) to eliminate defects in the processes at their source or at least to better manage those defects; and (4) to use teams to apply the tools. Still, the question on the table—which tool was best under what circumstance?

Perhaps more importantly, there were other issues that needed to be addressed beyond the tools and strategies. A concern for the steer-ing team was the fact that some studies suggested that cost cutting did not work very well and often was not sustainable. As one of them suggested, our costs are the consequence of our business sys-tem design, our processes and practices, and we need to improve those. It also seemed that the businesses and plants that made the most progress had better "leadership," better alignment to their goals, better teamwork, and more disciplined performance measure-ments, and somehow they had managed to get their people engaged in changing the organization. What were the processes that needed to be put in place to address these fundamental issues? Bob felt that the leadership of the organization must provide clear direction as to the strategy, the tools, and the goals, to assure its alignment to those practices. Otherwise, the organization would get less-than-optimal results. In fact, they could be counterproductive. The importance of organizational alignment is discussed in the next chapter.

References

1. Moore, R., *Making Common Sense Common Practice: Models for Manufacturing Excellence*, Boston, MA and London, England: Elsevier Butterworth-Heinemann, 2004.

2. Treacy, M., *Double Digit Growth*, New York: The Penguin Group, 2003.

3. Christopher, M., *Logistics and Supply Chain Management*, London: Financial Times/Prentice Hall Pearson Education, 1998.

4. Bossidy, L., and Charan, R., *Confronting Reality*, New York: Crown Business, 2004.

5. Womack, J., Jones, D.T., Roos, D., *The Machine That Changed the World: The Story of Lean Production*, New York: Harper-Collins, 1991.

6. Standard, C., Davis, D., *Running Today's Factory: A Proven Strategy for Lean Manufacturing*, Cincinnati: Hanser Gardner Publications, 1999.

7. Liker, J., *The Toyota Way*, New York: McGraw-Hill, 2004.

8. Lavelle, L., "Corporate Liposuction Can Have Nasty Effects," *Business Week*, July 17, 2000.

9. Wysocki, B., Jr., "Lean—and Frail: Some Companies Cut Costs Too Far, Suffer 'Corporate Anorexia,'" *The Wall Street Journal*, July 5, 1995.

10. U.S. Conference Board Report, 1996.

11. Morris, J.R., Cascio, W.F., and Young, C.E., *Organizational Dynamic Newsletter*, Winter, 2000.

12. Imai, M., *Gemba Kaizen*, New York: McGraw-Hill, 1997.

13. Nakajima, S., *Total Productive Maintenance*, Portland, OR: Productivity Press, 1993.

14. Pande, P., Neuman, R., Cavanaugh, R., *The Six Sigma Way*, New York: McGraw-Hill, 2000.

15. Parker, R., and Carlisle, J., *Beyond Negotiation-Redeeming Customer-Supplier Relationships*, New York: John Wiley & Sons, 1991.

16. Moubray, J., *Reliability Centered Maintenance - RCM II*, New York: Industrial Press, 1997.

17. *National Center for Manufacturing Science Newsletter*, Ann Arbor, MI, November, 1991.

18. *The Benchmarking Management Guide*, Cambridge, MA: Productivity Press, 1993.

Aligning the Organization

2

Management has no more critical role than to motivate and engage large numbers of people to work together toward a common goal... assisting them by removing obstacles (to their success).

Gary Convis, President, Toyota Motor Manufacturing, USA

Bob Neurath agrees with Gary Convis about the criticality of aligning the organization. He understands that it's common to hear the following phrases within an organization and agrees with them: "We've got to get everybody pulling on the oars at the same time," or "We've got to get everybody on board with our strategy." Maybe the need for these kinds of rhetorical statements is obvious, but just exactly *why* is it so important to *align an organization to a common strategy and set of goals?* What are the fundamental drivers for this need, and how do we accomplish that elusive alignment that is so badly needed? This need will be discussed in this chapter.

Why We Must Align the Organization

In Edgar Schein's groundbreaking book, *Organizational Psychology,*[1] he opines that:

> "The first major problem of groups in organizations is how to make them effective in fulfilling both organizational goals *and* the needs of their members. The second major problem is how to establish condi-

tions *between* groups which will enhance the productivity of each without destroying intergroup relations and coordination. This problem exists because as groups become more committed to their own goals and norms, they are likely to become competitive with one another and seek to undermine their rivals' activities, thereby becoming a liability to the organization as a whole. The overall problem, then, is how to establish collaborative intergroup relations in *those situations where task interdependence or the need for unity makes collaboration a necessary prerequisite for organizational effectiveness.*" (emphasis added)

This seems to be a reasonable reflection of reality. Given this, one should be especially careful about using Management by Objectives (MBO), especially when the objectives and their measures are not fully aligned. This is particularly true when there is a high degree of interdependence between groups, *but* the groups are held accountable for *different* and sometimes oppositional measures. Two groups that clearly fit this description are production and maintenance in a manufacturing plant. The opportunity for conflict is very high, increasing the likelihood for poorer overall performance, despite clear individual group objectives. For example, in a manufacturing plant, it is common for production to have specific production and delivery date targets. It is also common for the maintenance staff to have specific spending budgets they must meet. What happens when these goals conflict? Suppose production decides to run the equipment above its inherent capability to meet its targets, inducing defects and failures in the equipment and additional maintenance costs beyond the control of maintenance. Or, suppose maintenance decides to keep a set of equipment down longer to break its preventive maintenance (PM) compliance record resulting in late shipments to a customer. These are competing objectives, one of which is to the detriment of the other. It's essential that the competing groups have overarching goals that transcend any one group's goals.

Schein goes on to observe that the very process of organizing in order to better manage inherently creates competing groups—shifts, areas, departments, divisions, plants, and so on. And group competition between these functions is intrinsic in any complex organization. Each group "competes" for limited resources, recognition and rewards, executive attention, pay and bonuses, and so on. While organizing is essential for becoming more effective, a poor structure and/or strategy can be very detrimental to the overall goals. Group competition within an organization leads organizations to become "politicized" to one degree or another. Hence, we often hear a phrase similar to "it's just politics," when one person or group leader in the organization does something that is not well received in another group. Politics is simply the exercise of power to achieve a goal. When goals are not aligned, we can have destructive politics, which is not good for the overall health of the business.

Schein's studies also provide excellent insight into what happens both *within* and *between* competing groups. *Within* a group, its members become more closely knit, more task-oriented, and more formalized and structured to present a common position on most issues. On the other hand, in *between* group situations, each group tends to perceive the other groups as the "enemy," competing for recognition and rewards, scarce resources, executive attention, and seeking primarily to "win" the perceived prize. As subtle hostility increases toward other competing groups, perceptions are often distorted to support each group's own particular position. Examples of where this can happen in a manufacturing plant include competition between shifts, plants, production and maintenance, and even divisions.

Schein's studies also indicate that at any given time, you're likely to have a "winning" group, which tends to retain cohesion, but after the "win," the group becomes less task-oriented and more complacent. The "losing" group tends to rationalize its loss (e.g., the other side had an unfair advantage), but after losing, the group often becomes more intense and even more task-oriented and often reorganizes to become more effective and increase the probability of "winning." Thus, the cycle of competition among the groups continues. Some competition is healthy and creates a zeal for improvement, but excessive or *destructive* competition is not, particularly when it is covertly to the detriment of the organization's overall performance. Examples of destructive competition include running a process above its inherent capability to fulfill production quota, not sharing information across shifts to help all shifts, or not sharing an improved or recognized set of best practices among all plants. The list goes on, but the point is to focus on the *success of the businesses* more so than any one group or individual, while still recognizing the individual and/or group contributions. Success is always a shared responsibility and a shared result.

Since we can't eliminate shifts, departments, divisions, or other inherently competitive groups, how do we effectively use groups (or teams) to be more successful as a business? **We must align the organization to a higher purpose**, or what Schein calls superordinate goals—goals that have a higher purpose than any one group and align all groups to that purpose giving each a common goal and sense of identity. More importantly, whenever there is a conflict between groups, each must always revert to the superordinate goal/measures in making decisions. For example, "We're going to be the best manufacturing plant in the world," might be a superordinate goal. Or, constantly reminding people, especially managers, that we must always be asking "What's the right thing to do for the business overall, all things considered?" And, it's not enough to say any of these things once and expect people to respond. They must be constantly repeated until they become a habit or part of the culture of the organization. Some experts reportedly say that any particular alignment message must be repeated at least 21 times, each subsequent time proximate to the previous time, before it becomes embedded into the culture of the organization.

Aligning for Manufacturing Excellence

So, what specific steps should we take to align the organization? Below are several suggestions for aligning your organization for manufacturing excellence. These have been used in one form or another at Beta International to better align their organization. You may not need to apply all these, but they should help start the thinking process for better alignment and improved performance, or they may trigger other ideas that are more suited to your organization. Before you consider these, constantly keep in mind that people will almost always act in their personal interest. It's essential then that corporate interests and personal interests be sufficiently aligned so that both individuals and the corporation will benefit from that alignment and the resulting improved performance.

At the highest level, we might consider the model in Figure 2-1, one used by Beta to align the marketing and sales, manufacturing, and R&D functions. Each functional area has certain goals, e.g., marketing and sales for market share and booking volume, manufacturing for production, unit cost, and on-time delivery, and R&D for research into new products and processes. Each will be held accountable for

Figure 2-1. *Aligning Functional Elements to Superordinate Goals.*

its function's performance measures, and there might be common goals and measures between any two functional groups, but *all* will be held accountable for the corporation's superordinate goals, e.g., market share, return on assets, gross profit per unit of product, business volume percentage of new products, and so on. So, for example, if marketing meets its goal for booking volume, but does so through lower margin/higher cost products in manufacturing, the superordinate goals will not be met. If manufacturing meets its goal for unit cost, but does so through poorer on-time delivery performance, the goals will not be met. Each functional unit must ask two questions. Will my actions support my functional goals? And more importantly, will my actions support the corporation's goals? The superordinate goals always take precedence, and optimizing tradeoffs must be made regarding their achievement.

A particular focus is given below to internally aligning the manufacturing function. The suggestions provided could be used as a model for the internal alignment of other functions as well. Key suggestions are[2]:

- Communicate a common strategy with common goals and respective roles
- View safety and manufacturing excellence as an integrated process

- Foster a partnership for manufacturing excellence between production and maintenance
- Pause and reflect regularly
- Engage the shop floor in improving productivity
- Establish a manufacturing excellence forum
- Develop a set of questions for executives to ask when visiting plants to reinforce these principles

Each of these is discussed below.

Communicate a common strategy with common goals and respective roles. At the risk of stating the obvious (although often not achieved), if we are to align the organization to a common strategy and set of goals, then we must actually *have* a clear strategy and set of goals. As Schein observed, there should be overarching superordinate goals that embody our overall objectives. A word of caution here is that superordinate goals must be believable and achievable in a definitive time frame, even if that time period is several years. Otherwise it lacks credibility and can actually be a demotivator. Further, the leadership must constantly communicate, reinforce, reward, and then communicate again and again (21 times?) the strategy and goals, until they become embedded into the culture of the organization. It's also essential that we take every opportunity to articulate respective roles of members of the organization, while simultaneously reinforcing the need for teamwork in achieving the overall or superordinate organizational goals. In doing this, we must regularly foster frequent communication and interaction among the groups to assure they *identify with each other and with our common goals.* We may also need to occasionally rotate members of the various groups and use cross-functional teams, again to assure that everyone identifies with the greater purpose communicated in the superordinate goals.

View safety and manufacturing excellence as an integrated process. To their credit, most organizations place a huge emphasis on improving safety performance. And yet, their emphasis on best practice and manufacturing excellence is often less than emphatic. Studies have demonstrated clearly that safety and manufacturing excellence go hand in hand. For example, Overall Equipment Effectiveness (OEE) (to be reviewed in more detail later) and injury rate are inversely correlated, that is, the higher the OEE, the lower the injury rate. Correlation coefficients range from 60% to 90%.[2]

The more reliable and disciplined you are about operating your plant, the safer you'll be. In part this is because the plant "falls over" less frequently, requiring less intervention and a decreased risk of injury. In addition, plant discipline improves across all areas—operating, maintenance, and safety practices. If safety is not an option, then manufacturing excellence and best practice should not be an option either; they go hand in hand. A good news note for those who have been through a long, successful, safety improvement initiative—you can implement the same managerial steps used to improve safety performance to increase manufacturing performance. It's typical that a safety improvement program will have the following elements[3]:

- Top-down leadership—clear, consistent expectations
- Bottom-up ownership and employee engagement
- Education and training
- Action plans and measures
- Visual communication
- Standards and procedures
- Benchmarking and aggressive goals
- Audits and assessments
- Root cause focus—eliminate repeat failures
- Rewards (and "punishment" for failure to achieve desired goals)
- Resources for supporting improvement
- Continuous improvement expectation and process
- A culture … a way of life

These same elements apply to improving manufacturing performance.

Foster a partnership for manufacturing excellence between production and maintenance. The relationship between production and maintenance is often strained. As Winston Ledet observed, they often have opposing goals and measures. A common refrain from operations in many manufacturing plants is "we break 'em, you fix 'em," or from maintenance vice versa—"you break 'em, we fix 'em." A partnership agreement should be developed and documented that defines the relationship in general terms and provides measures that demonstrate that the partnership is working. Indeed, production and maintenance should share several *common* measures, for the partnership to be meaningful and lasting. For example, having production (and maintenance) responsible for measures such as down time, maintenance costs, and PM compliance would give a new perspective to production on these measures. Studies have shown that most equipment downtime is induced by poor operating practice, so *not* having production responsible for these key measures would simply be wrong.[2] Conversely, having maintenance responsible for on-time delivery would provide them with a new perspective on working with production to keep the equipment running. Other actions that we might take include linking various groups to common goals and measures. For example, having a percentage of each shift's performance measured according to the next shift's success, e.g., 30% of the 1st shift's performance is based on how well the 2nd shift does, 30% of the 2nd shift's performance is based on how well the 3rd shift does, and so on. With this approach, excellence in shift handover communication becomes essential. Similarly, between production areas, 30% of each area's performance might be based on how well the successive area does. Getting each group concerned about supporting the other groups within the organization improves *overall* organizational performance, not just that of a single group. The competition is *not* inside the walls of our company; it is outside, and often in a foreign country.

Pause and reflect regularly. The management team should have at least one "away day" per quarter to review its strategy, progress, successes, failures, and so on. Pausing to reflect, analyze, and modify its plans is part of the ongoing process of learning and improving. And these "away days" are opportune times to present the best practices, recognize achievements, and remove any potential roadblocks. Pausing to informally reflect, analyze, and improve on a daily basis, from the boardroom to the lunch room, is essential.

Engage the shop floor in improving productivity. Nothing changes until the shop floor does things differently (better). The shop floor knows more about the details of how things actually work on the floor than anyone else. So, we must capture and apply the tacit knowledge they possess to improve. When on the shop floor, managers should routinely share key metrics with people about cost, safety, training, reliability, and productivity. A word loosely used for engaging people in the improvement process is empowerment. We can begin our empowerment efforts by having a process for *routine structured improvement time* for cross-functional teams. Truly empowering people doing the work means they have[2]:

- Clear purpose and direction—a sense of meaning aligned to corporate vision/direction
- Boundaries for the empowerment given and considerable self-determination within those boundaries
- Openness and teamwork when doing the work
- Skills to operate within the boundaries, or any related training needs
- Measures of effectiveness of the empowerment—its impact on the business
- Continuing feedback and support from management, and
- Flexibility to address the changing needs that may affect the boundaries, training, measures, and so on.

Are the people on your shop floor empowered to make decisions affecting quality, equipment reliability, and process stability? Do they feel they are a part of the change process, or do they feel ignored by it? Empowering people using a proven model is much more likely to be effective and improve overall business performance. As Margaret Wheatley said, "People own what they create," so let them "create," that is, be a part of offering their ideas and suggestions on the changes that need to be made. Put another way, people don't resist change; they resist being changed. As we'll see in the chapter on Kaizen, engaging the shop floor in improvement is a top priority. This means that management must "go and see" at the shop floor level to truly understand the nature of the problems and opportunities that people face every day.

Establish a manufacturing excellence forum. This forum is typically a diagonal slice of the organization from the plant manager to the shop floor. It is both cross-functional and multi-layered but should not consist of more than 10 people. Its primary goal is to communicate the best practices across management layers and group boundaries, and to identify any obstacles to their implementation. Members should be rotated about once per year to keep the improvement ideas fresh.

Develop a set of questions for executives to ask when visiting plants to reinforce these principles. Executive leadership is paramount for achieving manufacturing excellence, and there is much room for improvement here. One straightforward approach for this leadership is getting executives to ask the right questions to better align the organization. These questions will vary from business to business but should constantly reinforce our strategy and goals using our super-ordinate goals to align the business to a higher purpose. What gets asked about usually gets done, so it's critical that executives ask the right questions. Table 2-1 provides a sampling of potential questions for executives to ask when visiting plants.

Developing and Using an Asset Management Strategy as an Alignment Tool

Over the past several years considerable progress has been made at Beta in increasing senior management awareness of the importance of reliability and manufacturing excellence principles, a condition that is difficult to find in most manufacturing plants. But this progress is not good enough. Without additional progress, jobs may continue to move to countries with lower labor costs; the sad thing is that it's just not necessary. The U.S. and so-called "Western" manufacturers *can* compete with the low labor costs of Asian and other foreign manufacturers, particularly those that are less labor intensive. A case study will be provided in Chapter 3 on this under the heading "Offshoring/Outsoucing." Moreover, consider the gains in market share by Toyota, Honda, and others that are making cars in the U.S. using essentially the same workforce and technology. The issue is NOT workforce capability or technology. It is management. U.S. managers, Beta's included, are often much too focused on costs and cost cutting and not enough on processes and overall performance. Costs are a consequence of your business system design, and market share and output are more important than costs per se, we've seen lots of studies indicate that cost cutting has a very low probability of success. Consider General Motors and Ford, who have both done extraordinary cost cutting over the past two decades, all the while continuing to lose market share and experience declines in overall business performance. Layoffs and cost cutting, while on rare occasion may be necessary, are not the way to inspire people to help improve productivity and the overall performance of the company—quite the contrary. The leverage to our success lies in getting hundreds, even thousands, of employees looking forward to coming into work and doing that work just a little bit better *every day*.

Table 2-1 *Potential Questions for Executives to Ask at Manufacturing Plants*

- Is our Overall Equipment Effectiveness (OEE) or Asset Utilization (AU) program in place down to the shop floor level for identification of all losses from ideal production? What's the value of the lost profit opportunity in the losses from ideal in our OEE/AU measurement? Are we closing the gap? How?

- Does the shop floor "own" and work to improve all the losses from ideal production? Do we have a prioritization process for improving these that considers gross profit contribution as a key consideration?

- What is our production schedule compliance? What is our daily production variability—average and standard deviation? What key steps are being taken to minimize variability and improve compliance to schedule? Are we using statistical analysis to set our inventory levels, especially between internal processes, so that we're confident of meeting demand, while maintaining minimum inventory?

- What is our inventory turns ratio on total inventory (not just finished goods)? What's our plan for reducing inventory while still retaining high on-time delivery?

- What is the relationship between operations and maintenance? Do cross-functional teams exist at the shop floor? Do they understand the cost of an hour of downtime and profit per pound after fixed costs are removed? Does production have and share ownership with maintenance for the success of the reliability effort?

- Is there a quantified critical equipment list for the unit? What percent of the equipment is critical? How is the list used to prioritize work?

- What is the Mean Time Between Failure (MTBF) of major equipment types (e.g., pumps, blowers, agitators, compressors, utilities)? Is it improving?

- What percent of time is the equipment available? What are the major causes of downtime? Is there an understanding of the impact of operational discipline on equipment breakdown? What is the relative magnitude of operational and design issues compared to mechanical causes of breakdown? What are the goals for improvement? Is there a plan with timelines and champions?

- What is the percent of planned and scheduled work? What is the percent of emergency and level 1 priority work orders?

Continued

Table 2-1 *Continued*

- What is the percent of PM and PdM work orders? Is it increasing?
- Where is most of the money spent in maintenance (pumps, piping, instruments, and so on)? Are these areas being targeted for cost reduction?
- What is the inherent reliability of the system? What does this mean in terms of maximum capacity? Does a reliability model exist to calculate the impact of improvement proposals?

Note: Clearly, not all of these questions will be appropriate for all businesses. Hence, these need to be vetted and condensed to the critical few, and others may need to be added as appropriate.

One issue related to operational excellence that frequently arises is that of physical asset management. A number of articles in the past few years have covered this topic and have made excellent points on this vital function. But these articles are typically more related to maintenance management than to true asset management. Calling the process described "Asset Management" does lend it greater business credibility and therefore may inflict an important shift in thinking in the minds of those readers who are in management. However, these articles typically still miss several important points. Let's add these points to the discussion, and perhaps create some additional debate on what asset management really is.

First, if we're going to do a good job at asset management, then it's essential that we understand the business expectations for the assets, both now and into the future. And, it's essential that we align the organization to that asset management strategy. One of the first questions to ask when considering any company's asset management strategy is: What are the company's business expectations for each of the major assets it employs, this year, in five years, in 10 years? The answer to this can have a huge impact on the asset management strategy. For example, if our assets will close in one year, then operating in a "staged decay" basis may be appropriate. On the other hand, if the business is growing at 5% per year for the next 10 years, and perhaps even introducing new products, then that dictates an entirely different strategy, one focused on operational excellence. It's pretty typical that little or no mention is made in the asset management strategy of the business expectations for the assets in the future, either overall or by plant. This is a critical omission, one which does not lend itself to proper alignment of the organization to the business objectives and its related asset management strategy supporting those objectives.

Second, we need to understand the assets' current condition and capability relative to the business requirements today and in the future. Is it reasonable to expect the assets to meet our business expectations given their current condition and capability? Where are the key gaps in performance and related condition, and how do we address those gaps? For example, if our OEE is 55%, and we're expecting to grow at 5% per year for the next five years, then our demand requirements can be met by improving our operational performance to 75% without additional capital. But, if we've been at 55% for the past five years with little change, it's *not* reasonable to expect this, unless we have a major shift in our strategy and practices. It's essential that we define current asset performance in terms of several key high-level measures, such as OEE, *unit* cost of production, injury rate, return on capital, and so on, and then set goals for their improvement in order to determine if our asset management strategy is effective. Further, an assessment must be made of our assets' current condition and any gaps in capability identified relative to future expectations, and an action plan developed thereto. While a better job is done here, there is typically insufficient discussion regarding current asset performance and condition, or how we address the gaps, and then measure the effectiveness of our asset management strategy. Doing so can be an excellent tool for aligning the organization to those goals.

Third, little mention is made of the role of operations in managing the assets. Data from most companies indicate that maintenance typically only controls about 10% of production losses from the ideal.[2] More specifically, two-thirds of production losses have little to do with equipment, being related to issues such as changeovers, startup and shutdown, and rate and quality losses. Of the one-third that is equipment downtime related, it's common that two-thirds of that is induced by poor operating practices. So, maintenance only controls one-third of one-third of the assets' production performance, or 10%. Further, it's common to have 2 to 10 times as many operators as maintainers in a manufacturing plant. Given that this data is even remotely accurate, how can we have an effective asset management strategy if operations don't take the lead role in our asset management strategy? Much the same as we as individuals "own" the reliability of our cars, with maintenance in support of it, operations must lead and own the reliability of the physical assets and, hence, must take the lead role in the asset management strategy and its implementation. The asset management strategies are typically maintenance led, and are focused predominantly, sometimes exclusively, on maintenance issues. Clearly, having a good maintenance strategy and good practices are essential for a good asset management strategy. But, it's not nearly sufficient and without addressing operational issues will yield only minimal improvement. Having operations take the lead for the asset management strategy will further align the organization to that strategy.

Fourth, insufficient attention is given to life cycle cost principles in the asset management strategy. For example, the capability, reliability, and ease of asset management begins with the design, and with the comprehensive application of Failure Modes and Effects Analysis (FMEA) and Reliability-Centered Maintenance (RCM) principles in the design. And yet, the Society for Maintenance and Reliability Professionals reports that 86% of its members surveyed do *not* use life cycle cost principles in their capital projects. If we don't get this right, it will make managing our assets much more difficult for the life of the asset. And maintenance typically has little control over capital projects. Life cycle cost principles, and the use of RCM/FMEA principles in the design, must be made an integral part of our asset management strategy. If we fail there, our assets will never be truly reliable, will be much more difficult to manage, and will be much more costly. Applying these principles in the design, with the support of operations and maintenance, will help further align the organization to our asset management strategy.

Finally, we must ask, how do we plan to implement our asset management strategy? Is it a document intended to be used by maintenance for doing better maintenance? Or, is it a defining policy document that demonstrates how we're going to use the assets for overall business success—improved market share, improved gross profit, higher return on capital (less capital spent, lower costs, and higher market share)? Who will lead its implementation—operations or maintenance? How will we train our people in its principles and use? How will we know it is being read, and applied; what are the measures that indicate this? Is there a summary document that captures the key points we want to make and use to constantly remind *all* our employees of its importance? Doing all this will assure that we align the organization to our asset management strategy. If we don't do this, the strategy may end up on the dusty end of a little-used bookshelf.

Most asset management strategies do a reasonable job outlining a maintenance management strategy, e.g., work management, planning and scheduling, equipment histories, material management, classical maintenance measures, job and cost control, preventive and predictive maintenance practices, contractor management, and so on. However, they generally fall short in characterizing a good asset management strategy. For proper alignment of the organization, a number of issues must be more fully addressed to assure excellence in asset management—understanding the business expectations for the assets, the current capability and condition of the assets, the related gaps, the use of life cycle cost principles at the beginning of the asset's life, and the role of operations in leading the development and implementation of the asset management strategy.

CEO Compensation: Mis-Aligning the Organization

Most people would agree that getting employee personal interests aligned with corporate interests is a necessary element in aligning an organization and maximizing motivation of the employees. Most would also agree that fiduciaries of the corporation, like the CEO, must put corporate interests ahead of their personal interests. That's part of the responsibility of a corporate fiduciary. It's not a giant leap in logic to understand that CEO pay averaging as much as 600 times that of the average employee, and CEOs receiving pay raises of 10% to 25% or more, while the average employee gets 3% to 4%, is inherently contrary to the concept of aligning employee and corporate interests. There's something wrong about such an arrangement; the CEO is perceived as getting a huge payout at the expense of all the employees.

In Bob Neurath's mind, CEO pay has the potential for being a huge detractor relative to aligning the organization for greater success. There continues to be lots of public criticism about CEO pay in general, and he's concerned that it may soon be his pay in particular. The unions are especially critical of CEO pay. They're asking why they should give up benefits and potential pay, when the CEO is being rewarded handsomely, something they think is at the rank-and-file's expense. Why should they improve productivity when jobs are going overseas anyway? What compensation policies are needed at Beta for fair and equitable treatment of all employees, from the CEO to the shop floor? Without that, Bob worries that all the policies, practices, and speeches focused on aligning the organization will fall on deaf, disgruntled ears to the long-term detriment of the company. As Louis Uchitelle observed, something is wrong when general payrolls are anorexic and the CEO's pay is obese.[4] Employee personal interests must be aligned to corporate interests, and a condition where CEO pay is perceived as exceptionally unfair will have a strong deleterious effect on the alignment effort.

Bob Neurath is concerned about the impact CEO compensation will have on aligning the organization. Most of the articles that he has read suggest that exorbitant salaries appear common, along with the equally common rationalization that, to summarize, "they're worth it" in a free market. All the while, the rank and file grouse about frequent layoffs, outsourcing, and anemic pay raises. Just a sampling of the articles in his file regarding CEO pay over the past several years reveals the following:

CEOs of 50 companies who outsource the most work got bigger pay hikes—9% for them compared to 2% for their employees; their pay was also 28% higher than CEOs of the top 365 companies surveyed. CEO pay was 301 times higher than the average production worker, up from 42 times the average in 1982.[5]

It's the salaries of the CEOs that are too expensive and should be sent to other countries. According to Towers Perrin, U.S. CEOs earn 23 times as much as those in mainland China; 10 times those in India; 9 times those in Taiwan; 5 times those in Japan; 4 times those in Spain; 3 times those in the UK, France, Italy, and the Netherlands; and 2 times those in Germany and Switzerland; CEO salaries have climbed from $1.4M in 1980 to $8.1M in 2003, 8% per year compounded. Between 1980 and 2003, CEO salaries rose 480%, adjusted for inflation, while corporate profits rose 145%. Meanwhile worker productivity rose 61% and worker pay stalled. CEOs made 44 times as much as workers in 1980, and 254 times as much in 2003.[6]

CEOs pulled in median compensation of $14.1M in 2004, up 25% from 2003. Many boards try to keep CEO pay above the median, a practice known to critics as the Lake Wobegon effect, where all executives are above average.[7]

"Large pay packages continue to touch a raw nerve. As long as boards are unaccountable, Corporate America won't change and fundamental problems will remain."[8]

CEO pay and company performance are not correlated. Many CEOs get highly compensated for poor company performance, while many other CEOs are poorly compensated for excellent company performance.[9–13]

On review of the CEO pay of 77 of the top 100 British companies, the Financial Times concluded that "…there is no discernible relationship between CEO pay and their companies' performance." Further, Rich and Larson examined CEO compensation programs and shareholder returns in 90 U.S. companies to determine if top executive compensation plans made a difference. They were unable to find any difference. A 1994 study by the National Institute of Economic and Social

Research found that between 1985 and 1990 top executive pay increased by 77%, while earnings growth increased by 17%.[12]

Sixty of the worst performing companies in the Russell 3000 lost $769B in market value over five years, while the aggregate pay for the top five executives in those companies was $12B.[13]

Between 1994 and 2003 executive pay nearly tripled from $2.5M per year to $6.2M, an 11% compounded rate (while most employees were looking at 3% to 4%). Meanwhile shareholder return bounced around from stellar to lackluster to outright disastrous. For example, earnings in 2003 were about the same at 1997.[14]

Rank and file employees are expected to receive record low pay increases for 2004—3.3% says Hewitt Associates. But, CEO pay should be up 8.8% to a $7.3M average.[15]

CEO bonuses rose 46.4% at 100 big firms in 2004. CEO total direct compensation was 160 times the average production worker.[16]

In recent years there has been a huge increase in CEO compensation, with the average pay for CEOs of publicly traded companies now approaching 600 times that of the average company employee, a 15-fold increase since the 1960s.[17]

Executive pay moderated in 2004 as a result of the impact of corporate reform. A survey of 367 CEO pay packages showed that total CEO pay was up smartly, to an average $9.6M—a 15% increase from 2003. But that average was skewed by the outsize pay package of a single CEO. Take it out of the mix and the average raise was 11.3%, not far off the rise in shareholder gains.[18]

While the criticisms are numerous and the unions, as well as mutual funds, are getting increasingly hostile, there seem to be few guidelines on setting CEO pay, other than applying market rates, which seem to be spiraling upward, and suggesting that "the board should do something." But what? Beta is considering using the principle of "internal equitability and external competitiveness" for what the board should do to establish pay guidelines, ones that are fair to the CEO, and as importantly to the future of the business, are fair to the thousands of other company employees. The data from the aforementioned studies suggest that CEOs are making huge sums for mediocre performance, while the rank and file suffer the insecurity of layoffs, outsourcing, and pay constraints. Where is the equity in this for the employees? How does this sustain the performance of the company over the long term?

How can CEOs claim such a single-handed contribution to performance? Current CEO pay levels seem too high to be sustainable, and in the eyes of the employees and unions are often unconscionable.

It is doubtful that there's ever been a perfect compensation system. There are just too many variables that influence compensation, too many individual value systems that influence their sense of fairness, and too many differing expectations in a workforce of any size to make everyone happy with their compensation. But, what a company can do is: (1) make sure its policies are designed to reward the behaviors and performance the company claims it values; and (2) clearly communicate these policies, making them transparent, so that even if people don't agree with the company's policies, they can agree that they have been carefully considered and communicated and are being followed. Finally, a compensation system does not have to be complex to be effective; rather, the reverse should be true. The principle that compensation must be **"internally equitable and externally competitive"** serves as a basis for developing policies and guidelines for executive pay. Using this simple principle is discussed below.

Internal Equitability

Internal equitability means that pay is fair in light of issues such as education, experience, responsibility, effort, contribution, and results. It does not mean you need an elaborate scoring system for the relevant issues. At first a point system might seem fairer, and some companies use such a system. However, the more complex you make any system, the more difficult it is to administer. Complexity doesn't necessarily add fairness and can increase the opportunity for inequity, perceived or otherwise, and more opportunity for "gaming" the system.

Suppose pay isn't perceived as internally equitable; what's likely to happen? Inequity reduces morale in any organization, but particularly when pay is the measure of equity. All organizations have people who complain routinely about their pay (among many things), making them, and those listening to them, less productive. Of course, fairness, like beauty, is in the eye of the beholder, and what seems fair to one may seem quite unfair to another. Totally eliminating that sense of inequity is highly unlikely, but having a clear policy that minimizes it can be accomplished with an open, transparent, well-communicated manner, one that we should constantly strive to achieve. With that background, let's turn our attention to CEO compensation with particular consideration being given to the principle of internal equitability. Being externally competitive will be considered later.

CEO Compensation and Internal Equity

The vast majority of people expect that people of higher rank, authority, or responsibility will be paid more. How much more? W. Edwards Deming suggested that once the CEO's compensation goes beyond approximately 20 times the wage of the average employee that a sense of *in*equity prevails. Warren E. Buffett has called executive pay the acid test of governance,[19] and "...often tells corporate chiefs to end practices ranging from huge CEO pay to incomprehensible financial reports."[20] The higher the gap between CEO pay and that of the worker, the greater the sense of inequity and unfairness that's likely to be present, reducing morale and productivity. And, the greater the inequity, the more militant unions will be. All this detracts from the success of the company.

As shown below, there continue to be huge increases in CEO compensation, with the average pay for CEOs of publicly traded companies somewhere between 160 and 600 times that of the average company employee, a huge increase over the rank and file pay. The thinking behind this shift in CEO compensation is apparently to provide incentives and rewards for CEOs that achieve exceptional results, those results being reflected in the company's stock price and shareholder wealth. However, this concept seems to ignore, or at least not fully recognize, a number of principles, all of which do need to be recognized, *as a matter of policy*, in setting CEO compensation:

1. First and foremost, as the lead representative of shareholders, CEOs have an inherent fiduciary duty to lead the company in achieving a high level of performance, period. Why should they be paid extravagantly for doing the job they were hired to do? How would the CEO respond to employees demanding significantly higher pay for doing what they were hired to do? Some have rationalized these exorbitant salaries as acceptable since they only amount to a small fraction of the market capitalization of the company. How would the CEO respond if we applied that logic to everyone's pay? We know how they respond to any extraordinary increases in costs.

2. Overall industry performance has not been used effectively as a normalizing factor. As the old saying goes, "Everybody's a genius in a bull market." Many of these compensation packages seem to ignore the need to normalize the company's performance to industry performance. If the company's industry shows strong annual growth, return on equity, profits, and so on, and the company is performing near the industry's average, then the CEO should *not* be rewarded handsomely for mediocre performance. The CEO's individual performance may have had little to do with the results achieved. Conversely, how often does one read of a CEO's pay being reduced if the company's performance is below the industry average? Some have suggested that this approach might lead to the CEO earning a huge bonus when the entire industry has had significant downturn, simply because they beat industry averages. This can be managed by having criteria that say if the company loses money, then no bonus is due. Life is tough like that sometimes.

3. As previously noted, several reviews show a poor correlation between CEO compensation and company performance. But, it should be correlated. The board of directors should be held accountable for assuring a strong relationship between pay and performance, particularly over the longer term.

4. Strategic issues and CEO pay should relate to the company's success in 3, 5, and 10 years. Bill Brenneman said that the ability to be strategic has to do with the capacity to understand the consequences of one's decisions over the time frame associated with their level of responsibility. Incentivizing CEOs with huge pay packages and stock options that vest in one to three years is *not* being strategic at their level of responsibility. There is a huge lag time in the response of big business "systems," and this response time is typically measured in many years, not many months. Generally, the bigger the business, the longer the response time.

5. The concept of being strategic also bears a need for fairness. Improved short-term results have often been at the expense of layoffs and low pay increases for the people who do the day-to-day work that makes the company more successful and the CEO wealthier. Layoffs, pay reductions, outsourcing, and longer hours create a perception of unfairness in the compensation system. It is impossible to persuade the rank and file that the CEO should benefit extravagantly by making them increasingly

stressed, or even unemployed. The shop floor will respond accordingly, although more subtly. The situation smacks of Marie Antoinette's attitude, who said of the starving masses, "Let them eat cake." We know what happened there.

6. Human nature is not taken into consideration. People below the CEO, particularly middle management and the shop floor, need to feel as if they are treated fairly. After all, nothing changes until the shop floor does things differently; these employees are performing the work that must be done for the company to implement its plans and achieve its goals. Every reasonable shop floor employee or middle manager has a clear understanding that CEOs need to be well paid. However, they also want to be treated fairly and rewarded for their contribution to the success of the company. Not surprisingly, they are particularly resentful of senior management being paid millions when the company is not doing well, and they are being asked to sacrifice. That resentment increases as the performance of the company decreases. Imagine the demotivating effect that this has on people, the company's performance, and its future.

7. The principles of internal equity and plain common sense have not been taken into account. Philip Adler said it best: "Where the differential pay between the top person's pay and that of the rank and file is perceived as unjustifiably large the consequences can be serious. Not only can it breed divisive class conflict and make wage restraint virtually impossible to achieve, it can also lead to a deterioration in the moral climate. If top salaries are seen to reflect greed and the abuse of power, an atmosphere can grow up in which fiddling of expenses and other dishonest practices become common place."[11]

    Studies also suggest that money is not a motivator. Being trusted to do a good job, doing the work you like, having control over it, taking pride in your work, feeling appreciated, and having dignity in your work—pride, enjoyment, and trust—are much stronger motivators.[21] That said, pay must still be internally equitable and externally competitive.

So, what's the right thing to do? Compensation for everyone, especially the CEO, must be internally equitable, considering these principles. The boards of directors must be held accountable for the application of these principles, assuring equity in the company's compensation system, providing a sense of alignment and teamwork throughout, and supporting the company's long-term success.

External Competitiveness

What happens if employees' pay is not externally competitive, that is, it is lower than the market rate? Beyond its immediate demoralizing effect on existing employees, ultimately many will leave to seek higher pay, leaving the company with a reduced ability to generate sales, revenues, profits, and a future. Replacing departed employees will become increasingly difficult, as candidates take themselves out of consideration after learning they can earn higher salaries elsewhere.

What happens if a company's compensation falls on the other side of being externally competitive? That is, it is higher than the market rate, considering productivity performance. It's much more likely that the company will not remain competitive, that profits will be inadequate to sustain the company's growth, and ultimately adjustments to pay, and possibly benefits, will have to be made. Employees will justifiably see such actions as "take-aways."

How do we determine what's externally competitive? It's typical to conduct a periodic survey of similar companies that hire the same types of employees. The goal is to compare positions with similar duties, responsibilities, and experience levels, and their corresponding pay and other compensation. Ideally, we should also include some measure(s) of employee productivity, or in the case of the CEO, company performance measures such as return on equity in their industry. Result-oriented productivity measures, such as revenue per employee and profit per employee are preferred. A company may be willing to pay more for a higher level of performance or may need to pay less for a lower level. A successful compensation system should show employees how the pay for each position compares to those at competing companies, and it should explain any discrepancies. For example, you may have a relatively young workforce and be growing at some 30% per year. The inexperience in the workforce could justify a lower pay, but strong growth offers future opportunity for rapid advancement and higher pay.

CEO Compensation and External Competitiveness

Many would argue that we need to be able to pay "top dollar" to attract top talent. While this may be true in theory, it seems more problematic in practice. As previously discussed, there have been several studies that indicate there is no discernible relationship between CEO pay and company performance; that top executive compensation made no difference in its effect on return to shareholders; and that top executive pay increased at a rate of more than four times that of earnings growth. Attracting top talent by using extravagant compensation packages does not appear to be providing top performance any more than it apparently does for professional sports teams. It seems that CEO pay is being fueled by a circular argument from those who benefit most from the argument; more and more CEOs being lured by higher and higher pay, justifying each increase by the most recent pay package for their peers. This circular argument is only exacerbated by the Lake Wobegon effect previously described, where all of the executives are above average. The board of directors needs to be held directly accountable for improving the relationship between company results and CEO compensation, particularly over the long term, all the while applying the principle of having both internal equitability *and* external competitiveness.

The sports star analogy is often made with regard to executive compensation. That is, sports stars command huge salaries commensurate with what the market will bear; so should CEOs, the argument goes. This argument is faulty. Sports stars don't have thousands working for them, or people who look to them as a role model for leadership and fairness! Sports stars are often focused on their personal success, more so than the team's success, demonstrating ego-maniacal behavior. Is this the model we want our CEOs to emulate? Is this the model of behavior that will inspire employees to make the company successful? Is this the model that encourages people to support the greater good for the company's success? Hardly. The CEO has a fiduciary duty to put the needs of the corporation first, above personal needs, and must lead by example, one that creates a sense of fairness, alignment, and teamwork throughout.

Finally, here are a couple of brief anecdotes that say much about the lack of leadership on the principles related to CEO pay. In an interview with the CEO of a major airline, when asked if his pay was too high, the CEO stated, "That's not for me to say."[22] While this response is fine in an interview, it is not the correct one regarding company policy. The CEO is the leader of the organization, and should lead, with the board's advice, in setting the *policies* that the company uses to compensate *all* employees, including the CEO. The board of directors should then use those policies and guidelines to set the CEO's pay. In another article, fat CEO payouts during mergers were questioned as to whose interest was being served. One CEO offered the rationale that "...to get ultimate shareholder value, you want a management team that's economically indifferent to whether or not they will have a job when the deal is done. This is the way to compensate them for putting the shareholders in front."[23] This seems self-serving. The CEO has a fiduciary duty to put the corporation and shareholder first in any event, period. Why should they be paid extravagantly to do the job they're hired to do? They certainly don't think employees should take such a position. A Gallup poll found that 90% of Americans felt that people running corporations could not be trusted to look after the interests of their employees. It also found that 43% of Americans believed senior executives were only in it for themselves. The figure was even higher in England where 95% felt that executives were only in it for themselves.[24] Such data likely surprises few, but with this lack of trust and perception of selfishness on the part of executives, it will be exceptionally difficult to lead and align any organization to a higher purpose.

We live in a global, capitalistic, and Darwinian economy. The intense competition for survival and prosperity dictates that wages be constrained, or that productivity improvement be substantially more than wage and other cost increases to assure continued competitive position. Those situations that lead to a dysfunctional and unsustainable outcome (e.g., extravagant pay for one person in an organization at the perceived expense of all others) must be avoided. Adam Smith, the father of free-market thinking, believed that free markets also required sympathy, or caring for others and sharing the gains, and that without that there would be a breakdown of the trust on which the market depends for its healthy operation; Darwinian thinking is associated with the processes that create self-sustaining ecological systems, not simply the survival of the powerful for a brief period.[25] Think of this as having equity and fairness in the pay structure so that the business system has a greater probability for survival and sustainability. The CEO has a fiduciary duty to diligently strive to represent the best interests of the company. It is not in the company's long-term interest to alienate the rank and file employees through inequitable compensation. To support this, the board of directors should establish policies that consider the following in setting CEO compensation:

The board of directors should use the principle of "internal equitability and external competitiveness" in establishing CEO pay and balance the two concepts appropriately. Thus, CEO compensation must be seen as equitable down through the ranks to the shop floor level. CEOs should not be paid extravagantly at the risk of alienating the rank and file and the future prosperity of the company. Employees will not perceive equitable treatment in an environment where, as Louis Uchitelle states, the general payroll is anorexic while CEO pay is obese.

Industry performance must be included as a normalizing factor in judging company performance and using that performance to judge CEO pay. For example, if the average return on equity in a given industry has risen from 10% to 15% because of increased demand and pricing power, then its performance is only average compared to peers. Hence, the CEO likely deserves no more than an average increase in pay.

Strategic issues and CEO pay must consider the company's success in 3 to 10 years. Being paid handsomely after major cost cutting and subsequent profit improvement, only to see the company's performance deteriorate in future years because of inadequate investment today, is not being strategic. Thus, a large portion of compensation must be deferred in recognition of the need to be strategic about today's decisions. Being strategic is having the capacity to understand the consequences of one's decisions over the time frame associated with their level of responsibility. The CEO has the longest time horizon in the company.

The board should be cautious about seeking the next "superstar." The data say the person selected isn't likely to be the next one, nor do we want the standards we use in selecting CEOs to be the same as those for sports stars. Sports stars don't have thousands of people working for them, seeking their leadership and fair treatment.

The board of directors must avoid the Lake Wobegon effect where all executives are above average. This only leads to ever-spiraling and unjustified increases in pay.

Finally, CEOs have a fiduciary duty to serve the corporation's interests, placing those interests above any personal interests. Why should they be paid extravagantly for doing the job they were hired to do, particularly if their pay is to the detriment of overall corporate morale?

CEO pay must not detract from aligning the organization, which would create distrust and discontent throughout. Indeed, CEO pay must follow the principle of being internally equitable and externally competitive. Beta will apply these principles to help better align the organization.

Initiative Overload: A Mis-Alignment Issue

A second mis-alignment issue is initiative overload—too many initiatives competing for limited resources and lacking in alignment to an overall strategy that links them in an understandable way. Beta is like most companies and has numerous initiatives. Each one is justified on its own merits and is typically approved as part of a strategy for improvement in that area. Examples include improving safety and environmental performance, driving major cost reductions, market expansion in product line and/or geographic areas, procurement initiatives for consolidation of suppliers or supply chain management, and often the use of one or more new tools for improvement (e.g., Six Sigma, TPM, RCM, and so on). Quality initiatives such as Total Quality Management (TQM) appear to have gone through their fad phase.

In the eyes of most people on the shop floor and in middle management, these initiatives appear to be independent and not linked to any particular overall strategy. The safety initiative is driven by the safety function, environmental by the environmental function, cost by the business unit leaders, procurement by the purchasing function, and so on. Each one is demanded as a matter of policy. Each initiative saps resources, which taken individually does not seem excessive, but taken in the aggregate very often is. Unfortunately, when middle management and the people on the floor try to implement these, there is often substantial dilution of their effort, or at worst confusion and failure. Most just try to do enough to satisfy the minimum requirement and not get into trouble. It's very difficult to have good organizational alignment when each of the functional groups is claiming priority without any particular overarching strategy that demonstrates how each fits into that strategy.

Bob Neurath believes Beta needs to do a much better job at this and plans to use the following approach:

> Get the basics right first—focus on reliability and stability as a foundational element of the strategy. If we get the basics right—excellence in operations and maintenance practices and good communication with marketing and R&D, and an overall strategy—we'll do better. We'll have greater reliability and stability in our processes, less variability, and lower cost. We'll also have better quality and delivery. And, we'll earn a bonus in fewer injuries and environmental incidents, and it will be easier to achieve the other initiatives. We won't have as many people scurrying about "putting out fires" and they'll be available to support higher value-add activities.
>
> Develop and communicate the company's strategy, including an explanation of how all the initiatives support that strategy, including a general priority for each.
>
> Identify specifically how each initiative will be resourced, particularly at the plants. Most of the initiatives converge at the plants. Having adequate resources to implement the initiatives and keep the plant running well is critical to the success of the initiative for the overall strategy.
>
> Keep the strategy and initiatives as simple as possible, letting the people on the floor work out the complexity in the context of their day-to-day operation.
>
> Communicate the strategy, its progress, and any clarifications regularly, leaving the communication channel open for criticism from the people actually implementing it and its related initiatives, so that adjustments can be made in a timely manner.

Summary

Peter Wickens said that one of the most important things he did as Managing Director of Nissan-United Kingdom, a very successful organization, was to align the organization to a common strategy and set of goals: "... *a leader (must) align the organization, to ensure that all its components are pointing in the same direction; that all are working together to achieve the same objectives."*[11] With all this in mind, more than rhetoric is needed to align the organization. There's ample evidence that it does work, but before launching into an "alignment" exercise, Bob still wants to make sure that the tools are well understood before further developing the strategy and aligning the company to that strategy.

More importantly, as CEO, Bob Neurath wants to understand all the other issues that will affect Beta's success. Beyond getting the marketing and sales right, as discussed in Chapter 1, for example, questions that keep going through his mind are:

> What about innovation? Innovate or die is an oft heard refrain. How will Beta position itself to continue to innovate and develop new products and processes, despite continuing price and cost pressures that sap away funding for that innovation?
>
> What about outsourcing, or off-shoring as it is sometimes called? Given the extreme price/cost pressures in some businesses, does it make sense to move some operations to foreign countries, where labor costs are much lower? What are the benefits? What are the risks? Does having a manufacturing presence in China or India offer strategic market or cost advantages, in those countries, or in other markets?
>
> How is Beta going to manage change to an even greater extent that it has already? Too many people seem to think that now that the company is back from the brink, things are fine. They're not, and continued and sometimes dramatic change and improvement are a must.
>
> What are the key leadership issues that need to be addressed? Is one leadership "style" better than another? Can leaders be made, or are they "born"? With so many books on leadership today, what are the "critical few" traits needed for effective leadership?
>
> What about teams? Teams and teamwork are touted as essential, and it seems to be a dogma that teams work, but do they really work? If so, under what circumstances? If not, under what circumstances? When and how do teams work best?

Clearly, there's plenty on this CEO's plate. The next chapters will explore those principles that Beta will use to address these critical issues.

References

1. Schein, E., *Organizational Psychology*, Englewood Cliffs, NJ: Prentice-Hall, 1980.

2. Moore, R., *Making Common Sense Common Practice: Models for Manufacturing Excellence*, Boston, MA and London, England: Elsevier Butterworth-Heinemann, 2004.

3. Leonard, J., Personal Communications, 2005.

4. Uchitelle, L., *The Disposable American—Layoffs and Their Consequences*, New York, NY: Alfred A. Knopf, 2006.

5. Hopkins, A., "Study Finds CEOs Who Outsource Get Bigger Pay Hikes," *Reuters*, August 31, 2005.

6. Sklar, H., "Keep Jobs Here, Send CEOs to Other Countries," *Lexington Herald Leader*, May 2, 2004.

7. Strauss, G., and Hansen, B., "CEO Pay Is 'Business as Usual,'" *USA Today*, March 31, 2005.

8. Bebchuk, L., *Pay Without Performance: The Unfulfilled Promise of Executive Compensation*. Cambridge, MA: Harvard University Press, 2004.

9. Iwata, E., and Hansen, B., "Pay, Performance Don't Always Add Up," *USA Today*, April 30, 2004.

10. Elliott, G., "What is Your CEO Worth?" *The Weekend Australian*, November 1–2, 2003.

11. Lavelle, L., "Executive Pay," *Business Week*, April 19, 2004

12. Wickens, P., *The Ascendant Organization*, London, England: MacMillan Press, Ltd., 1995.

13. Farrell, G., "Pension Funds Pin Target," *USA Today*, December 15, 2005.

14. Eisinger, J., "Ahead of the Tape," *The Wall Street Journal*, April 26, 2004.

15. Lavelle, L., "Another Stellar Year for Honchos," *Business Week*, December 27, 2004.

16. Lublin, J. S., "CEO Bonuses Rose 46.4% at 100 Big Firms," *The Wall Street Journal*, February 25, 2005.

17. "What We Learned In 2002," Editorial, p. 170, *Business Week*, December 30, 2002.

18. Lavelle, L., The 2005 Business Week Executive Compensation Scoreboard, 2005.

19. Lavelle, L., "Lessons of the 'Holliger Chronicles,'" *Business Week*, September 13, 2004.

20. Langley, M., "In Tough Times for CEOs, They Head to Warren Buffett's Table," *The Wall Street Journal*, November 14, 2003.

21. "Randstad North America and Roper Starch Worldwide, 2000," *Knoxville News Sentinel*, May 29, 2001.

22. Dixon, G., Quantas CEO, NBN TV Australia interview, August 22, 2004.

23. Thornton, E., "Fat Merger Payouts for CEOs," *Business Week*, December 12, 2005.

24. Handy, C., *What's a Business For?*, Boston, MA: Harvard Business School Publishing, 2002.

25. Burns, D., Personal Correspondence, Victoria, Australia, January 5, 2006.

Innovation

3

If you're not constantly improving, you're getting behind.

Introduction

The concept of innovation has existed for a long time. Webster's dictionary defines innovation as the introduction of something new—a new idea, method, or device. In most people's mind today, innovation, more often than not, conjures up images of the latest gizmo or some advanced "gee-whiz" technology.

One of the clarion calls in all businesses is to innovate, that is, they must develop or access the latest advance in technology, products, or processes in order to gain a competitive advantage. And yet, many companies are managing this process with only mediocre results. How can you achieve advances through innovation in a climate of mediocrity? In manufacturing companies, working to create innovative products without having your basic practices in order is problematic at best, and nearly impossible at worst. You might, by some accident or happen-stance, but innovation and the attitude and culture that goes with an innovative company must permeate the organization from the CEO *to the shop floor*. It's quite difficult to create an innovative environment in an organization when only one small group (e.g., research and development [R&D]) is perceived to be responsible for innovation.

In his book *Gemba Kaizen*,[1] Masaaki Imai observes that Western companies tend to worship innovation, where innovation is focused on major changes following technological breakthroughs, the latest management fad, or new production techniques. According to Imai, this is a high-risk approach to innovation. On the other hand, he suggests that Kaizen, or day-to-day improvement and innovation on the shop floor, is a more practical common sense approach that carries much less risk. Change is subtle and incremental. If you make a mistake, it's easy enough to correct it without having a major impact on the company. Perhaps more importantly, it creates an internal competitive advantage through the tacit knowledge of the thousands of little things that you do better than your competition. Bob Neurath's view is that we must do both! Consider Figure 3-1.[2]

Three companies are represented in this "market survivor profile." Each has a certain market share, and each produces its product for a certain unit cost. Each is likely to have a very different operating environment—C is compelled to change *to survive*; A is driven to change to *sustain and improve* its position; and B is typically relatively *complacent* with its position, at least in relation to A and C. After all, B is making money, has decent market share, and is working on its problems, but its sense of urgency is not like that of A or C, and unfortunately, it's more likely to become similar to a C before too long.

Each sells it products into a market place that has a variable market price. Market price depends on economic conditions, competitive pres-

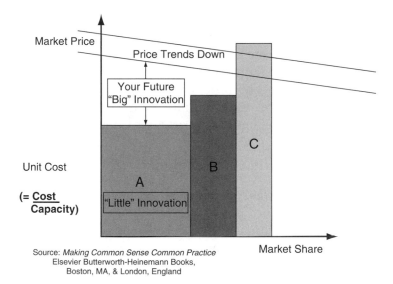

Figure 3-1   *Market Survivor Profile. We create a future by driving unit costs down because price trends down.*

sures, substitute products, and so on. Over time, for most products, market price tends to decline when adjusted for inflation. Over time, we must lower our costs to remain reasonably profitable, or we must create new products for which we can charge a higher price relative to older products. Even these new products will eventually become old and have diminished margins. So it's important in creating a future for the business that we constantly work to reduce our costs, allowing for adequate gross profits to finance future innovations for new products and markets. Why do prices tend to decline over time? A simplistic answer is that when two equivalent products for purchase are in front of us, we naturally tend to buy the cheaper one. We're trying to maximize our personal benefit with our limited resources. Manufacturers know this, and like Wal-Mart, they constantly work to put the cheapest products in front of us so we'll buy them, while they still make a profit.

As noted, the gap between unit cost of production and market price represents gross profit. How do we spend our gross profit? The gross profit finances R&D and new product/process development, marketing and sales and the development of our distribution channels, general and administrative expenses, and, of course, operating profits. These profits in turn finance additional capital investment and ultimately growth. Gross profit can be summed in one word—it represents the *future* of the company. Without gross profit for investing in the innovative efforts associated with marketing, R&D, and new product/process development, or what I'll call "big innovation," the company has little future. Of course, A has the brightest prospective future in this simple model. It has the greater ability to finance the "big innovation" required for assuring its future success.

How do we increasingly improve our gross profit and finance our "big innovation" in an intensely competitive world where price tends to trend downward? We must reduce our production unit costs, particularly when market price is declining with time, as they always do over the long term. How do we do that? By cost cutting? By process improvement? Both? As we've seen, cost cutting has a low probability for success. As we'll discuss below, it's through "little innovation" at the plant or operating level, constantly seeking to improve our processes so that the costs are constantly driven lower, and so that we can finance our "big innovation," perhaps allowing higher prices because of higher value in the product, allowing higher margins, and creating a virtuous circle for improvement. Imai would call this Kaizen, which we'll discuss more fully in later chapters. Let's now consider a few other issues.

Looking at Figure 3-1 again, let's consider the unit cost equation. It basically answers the following questions: How much did we make? How much did it cost? Thus, it provides a simple way of calculating our unit cost of production. At this point, however, let's make a modest adjustment in our thinking about this equation and view capacity as the maximum amount of product that could possibly be made with the current assets—no downtime, no rate losses, no quality losses, that is, zero losses from ideal. With that in mind, it's been my experience that the best companies focus on the denominator of the unit cost equation. That is, they focus on doing the things that will maximize the production capacity of their physical assets, and then they use that capacity to go after more market share, without having to make additional capital investments. Alternatively, with the capacity created (their hidden plant), they rationalize their worst performing assets. Or, they reduce their staffing levels for a given production capacity (e.g., reducing a six-day, three-shift operation to a six-day, two-shift operation). And, they work hard not to lay people off as a result of the rationalization. Rather, they re-deploy those people to different jobs that are more productive for the business or use attrition to manage the need for fewer people.

Clearly, if things aren't "falling over" because the processes are reliable and capable, the company will also have lower costs, not just because of potential volume, but also because costs just aren't being incurred unnecessarily. A word of caution—products are not made under the misguided guise of keeping unit costs down, thereby creating waste in the form of excess inventory and lower return on capital. Indeed, they are much more capable of running a lean manufacturing operation—their assets are reliable and capable of running when needed to meet market demand. They don't carry extra inventory, just in case, because they don't need to. Their assets don't fail when they're needed, and *all their costs are lower as a consequence of their business system design and capability.*

Cost Cutting vs. Creating an Innovative Environment

The data discussed in Chapter 1 indicate that cost cutting has a low probability of success in improving business performance. We may have to literally cut costs in the short term, just to survive, but it should not be our long-term strategy. Ask yourself how motivated you would personally be to innovate if you were working in a cost-cutting environment, where "Theory X" typically prevails (employees are lazy and need to be managed). This stands in sharp contrast to a "Theory Y" environment (employees want to make a difference and to be led). Most employees (95% or so) would much prefer to be led into making a difference and to take pride in their work and their company. We should *not* let the 5% that don't want to be led hinder the potential of all the rest. Indeed, we should sort them out.

As is noted in the previous text, *cost reduction is essential.* Cost reduction is differentiated from cost cutting in that it results from improving your processes to improve productivity; or eliminating defects to reduce failures; or applying a new technology to improve yields. Costs decline as a consequence. Costs cutting is simply reducing budgets, headcount, or other resources, and hoping that costs will decrease and overall performance will improve.

Getting Lower Costs Through "Little Innovation"

Production costs must be driven down to finance "big innovation," or step changes in technology or processes. The most effective approach for financing this is through "little innovation" at the plant and shop floor level, improving your processes so that the costs are not incurred, and eliminating the defects that cause the failures in the equipment and processes. Innovation here is about *everyone* constantly seeking to do the work a little better every day. Put the right processes in place, get your people engaged with a sense of ownership, create an environment for pride, enjoyment, and trust, and costs will decline as a consequence. Operating your assets in an optimal way will provide for minimum costs. This also creates an environment wherein the company is innovative from the CEO to the shop floor level, in all its key functions, and is constantly seeking to improve its performance in all areas.

There are lots of tools, methods, and strategies for achieving "little innovation," so that you can effectively finance "big innovation," and as importantly, so that you can create an environment for this "little innovation." As outlined in Chapter 1 and to be discussed in later chapters in more detail, these strategies include Lean Manufacturing, Kaizen, Six Sigma, Supply Chain Management, Reliability-Centered Maintenance, Total Productive Maintenance, Predictive Maintenance, Root Cause Analysis, and so on. Each of these tools provides a method for process improvement and cost reduction. The important point is that costs are reduced as a consequence of improving our processes and eliminating those defects that resulted in higher costs in the first place.

Creating Expectations

However, before we discuss the specifics of each of these tools, let's review what may be a more critical issue, that is, the concept of leaders engaging their entire workforce in helping the company to improve. There have been numerous studies that strongly support this as a key element in improvement, irrespective of the tools used. And, leaders who reorganize the business on a regular basis are more likely to add instability to their processes than improvement. Petronius Arbiter, in *Satyricon*, first century A.D., probably said it best:

> We trained hard, but it seemed that every time we were beginning to form up into teams we would be reorganized. I was to learn later in life that we tend to meet any new situation by reorganizing; and a wonderful method it can be for creating the illusion of progress while producing confusion, inefficiency, and demoralization.

This is not to say that we shouldn't occasionally reorganize to better align and manage the business to the tasks at hand, but it's also important to understand that each reorganization actually destabilizes the system for a period of time, often months.

The Hawthorne Effect[3] was one of the first studies of the importance of engaging people in the improvement process. As mentioned earlier, in the 1920s, Western Electric Company's Hawthorne, Illinois, manufacturing plant studied the effect of changing working conditions on productivity. For example, they speculated that if they improved the lighting in the plant, the workers would be more productive. They improved the lighting, and productivity improved. Then, to test their theory, the plant reversed the action, thinking productivity would decline. But, the opposite happened; they reduced the lighting and productivity improved even more. To summarize, in my words: if you give people a little attention, and they know you're expecting them to do better, they will. I think this principle always applies. No matter what tool you use, if people know you're looking for them to do better, they'll work hard with that tool to meet your expectations. This effect is similar to the so-called Pygmalion Effect described by J. Sterling Livingston,[4] that is, high expectations lead to the development of "super staff," whereas low expectations lead to an increased risk of failure. People in a subordinate position will work hard to meet your expectations, so long as the expectations are reasonable, and they have the time, tools, and training to do the work required. Similarly, Rosenthal found that expectations become a self-fulfilling prophecy. His research showed that teachers expecting better performance tend to get it, and likewise for coaches, researchers, and so on. Some 479 studies have found that teachers' expectations dramatically affect how students will do: high expectations yield better performance, improving student performance by as much as 30%, but as you might expect, low expectations yield poorer performance.[5]

In their book *How Full Is Your Bucket?*, Tom Rath and Donald O. Clifton[6] contend that praise is a powerful leadership strategy. Bad things can happen when people don't feel appreciated. Their research indicates that the workplace is most productive when the ratio of positive to negative interactions is 3:1. Things actually get worse if the ratio goes higher than 13:1. They also note that Gallup surveyed more than four million employees, finding that employees who give and get praise:

Increase their individual productivity

Increase engagement among their colleagues

Are more likely to stay with their current company

Receive higher loyalty and satisfaction scores from customers

Have better job safety records and fewer accidents on the job

So, why do some companies continue with a basic philosophy something akin to "the beatings will continue until morale improves"? Companies should work hard not to convey that impression and should give much greater weight to the positive rather than the negative. If you believe the work of Rath and Clifton above, that ratio should be between 3:1 and 13:1.

Csikszentmihalyi[7] makes the point that leadership and the creation of meaning at work demands that a business provides a better outcome for individuals in their daily working life, for society and, importantly, for the business in terms of sustainable performance. Leadership and creating meaning at work are in some ways like an exhortation for business leadership to create an environment where people see their work as "building something greater than themselves, a cathedral if you will," rather than "slaving away all day just to break up rocks" (that go into the cathedral).

David Burns in Melbourne, Australia,[8] has developed a graph that captures the essence of these principles in a single picture as shown in Figure 3-2. It's fairly common to use the Pareto analysis to rank the opportunities within an organization and to focus on those with the highest value. Pareto's law states that performance will depend disproportionately on doing a few things really well. As shown in Figure 3-2 we would typically select the top three to five opportunities and organize an improvement team together to address these. In Chapter 5 we'll review a process, the Business Level Failure Modes and Effects Analysis (FMEA), for identifying these top few opportunities and selecting the tools for addressing them. Typical tools used in these teams include Six Sigma, Reliability-Centered Maintenance, Root Cause Analysis, and so on. But, Burns also notes rightly that the value of these, while concentrated in five or fewer "projects," is typically less than the sum of all the rest of the small opportunities in the organization. For example, Winston

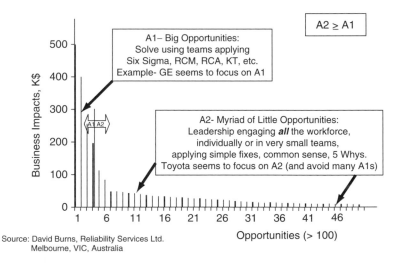

Source: David Burns, Reliability Services Ltd.
Melbourne, VIC, Australia

Figure 3–2   *Leaders Engaging the Entire Workforce in Improvement.*

Ledet suggests that it is common to have some 20,000 defects in a typical manufacturing plant, resulting in some 6,500 repair work orders, which can lead to 10 significant losses, which can lead to one major incident. Getting rid of all the little defects stops the bigger problems from happening. Similarly, Toyota is reported to expect that each employee will contribute some 70 improvement opportunities *per person per year*, most all of which are relatively minor, but which create enormous tacit knowledge and competitive advantage for the company.

Capturing those opportunities, however dilute they are, is essential if we are to sustain the improvement process and align and leverage the entire organization for superior results. This requires a more innovative approach to management, driving the responsibility for improvement to the shop floor. Not doing so ignores huge opportunity and results in small problems eventually becoming much bigger ones, both of which are to the detriment of the organization. A key message here is that there are the two broad categories of improvement opportunities, each with a major business impact. But, they require fundamentally different approaches for problem solving and implementation. Both should be taken seriously by management. Unfortunately many in management do not appear to sufficiently appreciate this phenomenon. It's essential to understand that a well-managed process for surfacing the many small scale opportunities and rapidly implementing them has a far greater impact on building and sustaining a continuous improvement culture in an organization than does a few large projects.

Finally, each Innovation, large or small, must be standardized so that the benefit is maintained within the organization. Hence, we need procedures, checklists, and so on, and people need to be trained in these and to practice these with great discipline. However, once the standard is set, we must be constantly looking for the next improvement. As we'll see when we cover Lean Manufacturing, innovation and standardization are two sides of the same coin.[9]

Outsourcing/Offshoring: The Loss of Innovation

Bob Neurath is supportive of these principles, but asks, what do we do in the interim while we get better alignment, cultural change, and productivity? Beta is under extreme cost/price pressure in some businesses, and it seems like the only way to address that is to move some of its operations offshore where labor rates are much lower. Indeed, Beta has many foreign operations including several in Asian countries. What are some of the considerations there?

Bob also reflects on the words of Alan Greenspan who said that original wealth comes from mining, agriculture, and manufacturing. Consistent with this, economists also report that for every job in manufacturing, there are three to five jobs elsewhere, in manufacturing support services, and particularly in local restaurants, hotels, car dealers, department stores, repair shops, and so on. If we continue to lose our competitive edge in those that create original wealth, we may lose our strong economic base because the entire economy is weaker. Offshoring may affect our economy in ways that over the long term weaken the entire economic base of the country. If we want to create original wealth here and build a strong economic base from which to assure a strong national economy, we need to try to avoid offshoring. At the same time, in some foreign markets, it may make sense to have a local manufacturing presence, particularly in large markets. This reduces shipping and attendant costs and risks and puts the manufacturing closer to the customer and their wants and needs. This makes the product more economical and available, increases profits and market share, and increases the opportunity for more innovation and market development.

Along this line of thinking, Bob also notes that labor is only about 15% to 20% of its total manufacturing cost base, so cutting labor costs by 20% will only reduce overall costs by 3% to 4%. And what about all the distribution and storage costs, the risk of supply disruptions to Beta's customers, the startup costs in a foreign country, and the political risks in a foreign country, particularly where labor costs are cheapest? This must all be considered.

With this in mind, Beta is like most companies looking to reduce its costs, particularly its labor costs, and has to consider moving its manufacturing operations to countries like Mexico and now China. Is this the right thing to do, and under what circumstances should we do this? What's the so-called "tipping point" at which it makes sense? Or alternatively, how productive do U.S. workers have to be to keep the manufacturing plants here? In 1926, Henry Ford remarked that "No greater injury can be done to a man than to pay him a high wage for a small amount of work, for then his high wage increases the prices of commodities and puts them beyond his reach." And, he also said, "If all the profits were given to the workers, then improvements... would not be possible. Prices would increase, and consumption would decline, and the business would gradually go out of existence. The profits have to go toward making lower costs, and the advantage of lower costs must be in large measure be given to the consumer. This in effect, is the same as raising wages."[10] Ford's words offer a bit of wisdom about capitalism. Let's look at some simple examples of these principles in practice. Let's try to answer one fundamental question: when does it make sense to "offshore" our manufacturing?

Two Scenarios

Let's suppose we have a U.S. or Western manufacturing company making and selling its products in the country of manufacture, and we have a foreign competitor making its products there, but selling them in the U.S. or Western country. Assume these two competitors have the cost profiles shown in Table 3-1.

There are several simplifying assumptions in these scenarios, so it's essential that Beta do the analysis for each business to see how these scenarios might more accurately compare. For example, shipping, import duties, and the risk/cost of capital for products in transit are higher for the foreign company, but most other costs are lower, e.g., manufacturing labor costs are much lower, as well as marketing, R&D, and general and administrative costs (G&A) and taxes. That is not to say that Asian companies aren't doing marketing or R&D, but they likely have lower labor costs for doing the equivalent effort.

In the first scenario, the Asian manufacturer is 130% more profitable than its Western counterpart. In this circumstance, Bob Neurath would have to represent the shareholder's interests first. If he knew Beta had to have Wal Mart's business to survive and prosper, where should Beta manufacture? Also note that literally cutting labor costs even by 33% (if that were even possible while still maintaining production) still leaves the Western business about 30% behind its Asian counterpart in profitability. In this scenario, cutting labor costs does not appear to be the answer. Moving Beta's manufacturing operations might be. But let's hold our judgment before making that decision.

Suppose in the above scenario that Beta's Western plant has mediocre manufacturing practices, a condition that is supported by a benchmarking exercise and other evidence. For example, in this situation the plant's overall equipment effectiveness (OEE) is 60%,

Table 3-1 *Comparative Costs and Gross Profits – Western v. Asian Company – Scenario 1*

	Western Co.	Asian Co.
Sales volume	$100	$100
Cost of goods manufactured		
Raw material	$ 35	$ 35
Energy and other variable costs	$ 15	$ 15
Labor	$ 15	$ 3
Depreciation and other costs	$ 5	$ 5
Other costs		
Packing and shipping	$ 2	$ 5
Cost of capital and risk, in transit	$ 1	$ 2
Import duties	$ 0	$ 5
Other costs, e.g., order processing, inventory carrying, commissions, pre- and post-sales service	$ 5	$ 5
Gross profit	$ 22	$ 25
Marketing and sales	$ 7	$ 5
R&D	$ 2	$ 1
G&A, interest and taxes	$ 7	$ 5
Net Profit	$ 6	$ 14

which is pretty typical in a discrete manufacturing plant and mediocre according to benchmarking standards. Suppose further that we could achieve a much higher level of performance, e.g., an OEE of 80%, but with 20% less labor, through manufacturing best practice, and we simultaneously improved our yields and energy efficiency just slightly, again through applying the right practices in a disciplined way. These are all achievable with the proper leadership, practices, and work culture. Suppose further that the market is expanding by 33% over the next few years, and that the additional business volume can be captured by *both* the Western plant and the Asian competitor. Then we might see the data displayed in Table 3-2.

At this point Beta's Western plant is competitive with its foreign counterpart, although the foreign counterpart is 33% more productive (with the same number of people). Note that Beta's competitive advantage is even greater if the foreign competition doesn't improve productivity at that rate. Note also that Beta's Western plant's labor costs are still four times higher than its competition, but it is competitive because labor is a relatively small portion of its total costs, and

Table 3-2 *Comparative Costs and Gross Profits: Western vs. Asian Company—Scenario 2*

	Western Co.	Asian Co.
Sales volume	$133	$133
Cost of goods manufactured		
Raw material	$ 45	$ 46.5
Energy and other variable costs	$ 19	$ 20
Labor	$ 12	$ 3
Depreciation and other costs	$ 5	$ 5
Other costs		
Packing and shipping	$ 2.7	$ 6.6
Cost of capital and risk, in transit	$ 1.3	$ 2.7
Import duties	$ 0	$ 6.6
Other costs, e.g., order processing, inventory carrying, commissions, pre- and post-sales service	$ 5.6	$ 5.6
Gross profit	$ 42.4	$ 37.0
Marketing and sales	$ 7	$ 5
R&D	$ 2	$ 1
G&A, interest and taxes	$ 7	$ 5
Net profit	$ 26.4	$ 26.0

it is much more productive. However, for it to be competitive has required that it become 66% more productive (33% more productive with 20% fewer people). This could also mean that Beta needs to shut down some operations (the poorest performing ones) that are no longer required to meet even an expanded market demand, taking advantage of its improved productivity. But, Beta would not need to move its operations, keeping those other additional three to five jobs in the community. While closing some poorly performing operations might be a bitter pill to swallow, it's better than the alternative—seeing the demise of the entire business unit, and *all* the jobs that go with it. At the same time, for new markets in foreign countries, consideration must be given to manufacturing in those countries where the products are sold, again making sure that costs are minimized and business results are maximized.

Many of Beta's businesses are already looking at these issues when considering offshoring, but Bob wants to make sure that this is being done in a comprehensive systematic way. So he instructs all his businesses to look at their businesses from this perspective—not just from a U.S. perspective—and also look at business units in Europe and Canada and ask the following: When comparing local and foreign production, where are the relative competitive advantages for each? Where is the "tipping point" at which it makes sense to retain the operation at its current location, including consideration of logistics costs and risks? What are the risks in moving? Or not moving? He asked that each major business unit report on this each year. More importantly, he instructed each business unit to take this information to the people in the plant, and make clear the business case for each plant. This was to be done along with providing the people with the tools, time, and training to allow them to achieve greater productivity and competitive position. And, he also instructed that if fewer people were needed as a result of improved productivity, that need would be managed through attrition, reduced overtime, reduced contracting, and employee transfers. Having been the victim of a layoff in his younger years and all the pain that accompanied it, Bob considered layoffs to be the last resort.

Human Capital: The Heart of Innovation and Value Creation[2]

This brings us to the next concern on Bob Neurath's mind. As noted, he was the victim of a layoff in his younger years, and it was a difficult time (several kids to care for, a mortgage to pay, car payments to make, other debts to settle, and limited funds to tend to all this). While he got through that period with only a few scars, he empathizes with people, perhaps more so than most, and feels a responsibility for their welfare in his decision making. At the same time, he knows he has a responsibility to the shareholders to make the business as successful as possible. Was his empathy for people getting in the way of making the difficult decisions to assure Beta's success? Why couldn't you have both—engaged people and a successful business? Wouldn't that actually improve the odds for the success of the business? He thinks yes, and that an engaged, highly motivated workforce will be much more innovative and productive at every level.

Bob is also concerned that all the experience in his current employees will be lost in the coming years as the baby boomers begin to retire. In some ways it's a blessing—a way to reduce staff as Beta improves productivity without having to impose layoffs. On the other hand, it has the potential to induce huge disruptions through the loss of a veteran workforce. What's all that human capital worth, he wonders? Bob is like most CEOs in that he often remarks that "our people are our most important asset," and his nature is not to dispose of assets, either human or physical, even in a downturn. He's now wondering what all those soon-to-retire assets are worth, and more importantly, what it might cost to replace them. To follow is a model for assessing their value.

Thomas Stewart and others have brought to the forefront the concept that intellectual or human capital has at least as much value as fixed asset capital. In his book "Intellectual Capital: The New Wealth of Organizations," he offers several models for understanding the value and impact of intellectual capital.[11] Recent trends in information technology also highlight companies like Microsoft that have a high degree of "intellectual capital," as perhaps being of greater value than the traditional capital intensive companies, like DuPont and others. Let's consider a simple way of thinking about, and methods for valuing, the acquisition (and "disposal") of human capital assets. The value of these assets, embodied by the knowledge, skills, experience, ability, and general intellectual capital of the workforce, must be recognized, if not in an accounting sense, at least in a managerial and decision making sense.

When companies acquire fixed assets, they generally follow a specific policy and process for the acquisition. Similarly, companies generally follow a specific policy and process for the acquisition of human assets. However, one of the key differences in the accounting treatment of these assets is that the costs and effort which goes into adding *fixed assets* are *capitalized*, and then carried in financial statements as *assets* that *depreciate* over time. Whereas the costs and effort for the acquisition of *human assets* are *expensed* and are not carried on the books as assets, not withstanding that they *appreciate* over time, especially in organizations that facilitate a continuous improvement and learning environment. The accounting treatment of these assets has a substantially different effect on the balance sheet, although both represent the investment of capital for sustaining and growing the business. And, perhaps more importantly, this accounting treatment has a substantial impact on the thinking of management. On the one hand, fixed assets are viewed as just that, assets on the balance sheet. On the other hand, people are often viewed as expenses, which can be more easily acquired and disposed. This can result in poorer decisions and may be fundamentally flawed.

Suppose you were a board member of Beta International and your CEO, Bob Neurath, announced at a board meeting that a major plant, one in which the company had invested $100s of millions, was being disposed, resulting in a major charge to the company. How would you feel about his judgment? On the other hand, if these were human assets being disposed through layoffs, how would you feel? If you're like most, there seems to be a tendency to place a greater sense of loss on the disposal of fixed assets than there does on the "disposal" of people, perhaps because of the belief that a greater cost is eliminated with the people. Whether this is appropriate or not can vary with the business circumstance. The value of human capital should be more fully accounted for when making these decisions. And, the accounting practices currently employed by most companies can have undue influence in driving the strategic decisions of many organizations. Fixed assets are treated as *assets that depreciate* over time. But, employees are treated as *expenses*, yet they *appreciate* over time with additional training and experience. The accounting of people should be such that their value as a corporate asset that appreciates over time is properly treated.

As we observed in Chapter 1, many studies indicate that cost cutting through layoffs has a high risk and a low probability of success. David Stamps also observed that downsizing in many cases is driven by a short-sighted, bean-counting mentality rather than a … strategic, long-term vision required to reshape a company into a smaller but healthier organization. Wall Street often rewards large layoffs with a jump in stock price; cost cutting must be good, since it will improve the bottom line. However, the evidence suggests that cost cutting is a poor bet, no matter how it is presented (e.g., downsizing, right sizing, restructuring, and so on). This is not to say that cost cutting never applies. Not all the companies studied suffered as a result of their cost cutting. Again, it may apply, for example, if you're on the verge of collapse, and have no choice to survive; if you're a bloated bureaucracy, and must do so to assure your longer-term success; or if you're faced with intransigence in unions or employees and need to get people's attention; in specifically targeted situations of obvious waste; or, finally, in a major market downturn or situation of global oversupply that severely depresses prices.

Let's return to the issue of valuing Beta's assets. Since Bob Neurath often refers to Beta's people as its most valuable asset, how might Beta assign a real value to those assets? At a very simple level we could apply a capital project model, that is, the same one used in acquiring fixed assets.

There are direct parallels between the process of acquiring an employee (a human capital asset) and that of acquiring a fixed capital asset. Let's consider Table 3-3, which shows the typical steps for acquiring these assets.

In acquiring a fixed asset, the company first identifies a need and then writes a specification. Then it issues a request for quotation and invites interested suppliers to bid. After evaluating the bids, the company selects the supplier. The equipment is delivered and installed. The company trains its workforce on how to operate, troubleshoot, and maintain the equipment. The equipment is started and production begins. If all goes smoothly, after a period of time the asset will be fully functional, and the company will have met its need by applying capital in an effective manner (e.g., expanding capacity, solving a production problem, addressing an environmental requirement, and so on).

Other considerations in the process include: (1) the capital expenditure being in line with the company's strategic plans for new products,

Table 3-3 *Process for Asset Acquisition*

Acquiring a Fixed Capital Asset	Acquiring a Human/Intellectual Capital Asset
The company:	The company:
1. Identifies a need	Identifies a need
2. Prepares a specification	Prepares a job description
3. Issues a request for bids	Begins recruiting activities
4. Evaluates the bids	Interviews and evaluates candidates
5. Selects a supplier	Makes a job offer; receives an acceptance
6. Takes delivery of equipment	Moves the employee to the company
7. Installs/starts up equipment	Indoctrinates/introduces the employee
8. Trains the workforce in its use	Trains the employee in specific job skills, and so on
9. Tolerates early low efficiency production	Tolerates early low efficiency production
10. Finally, works at full capacity	Finally, works at full capability

markets, and so on; (2) the expected rate of return on the investment; (3) the cash flow effects; (4) the cost of capital; (5) any fore-
gone opportunity for using the capital in some other manner; (6) the
anticipated life of the equipment; and (7) the tax consequences of the
capital purchase, and so on. It would be rare for a company to purchase fixed capital equipment without having a good estimate of its
costs and benefits.

As shown in Table 3-4, in acquiring human capital, most companies follow a similar process as that used for acquiring fixed capital.
Most are able to estimate the cost of the first five steps of Table 3-3 in
the process of acquiring those assets, even if they do not put the same

Table 3-4   *Valuation of Human/Intellectual Capital Asset Acquisition*

Task	Duration	Value/Cost
1. Identifies a need	1 day	$500/$260
2. Prepares a job description	1 day	$500/$260
3. Begins recruiting activities	1 day	$500/$260
4. Interviews and evaluates candidates	5 days	$2,500/1,300
5. Makes a job offer; receives acceptance	1 day	$500/$260
6. Moves employee to company location	$10,000	$10,000
7. Indoctrinates/introduces employee	10 days	$5,000/$2,600
8. Trains employee in job skills, on-job training, and so on	10 days	$5,000/$2,600
9. Employee performs at reduced rate	30 days@25%	$11,250/$5,850
10. Full productivity achieved, after	60 days@50%	$15,000/$7,800
Total Cost of Acquisition @ $500/day/@$260/day		**$50,750 / 31,190**

Note:
*1. In this example, two methods have been used to value the employee's time. The first
assumes a value of $400 per day ($100,000/yr per employee ÷ 250 days/yr). The second is
based on nominal salary plus fringes. For an employee making $50,000 per year, plus 30%
fringes, this would be $65,000/yr ÷ 250 days/yr = $260 per day. The actual calculation and
valuation process will of course vary from company to company and employee to employee.
The point is not to be precise about any particular value/cost of acquisition, but rather to
offer a model for the valuation.*

precision into the actions as they would for acquiring a fixed asset. The balance of the steps may be more difficult, but follow analogous steps for acquiring fixed assets. For example, the task of preparing the site for delivery and set up of the equipment is analogous to preparing the "site" for a new employee's arrival. The employee must have an office, a desk and chair, a telephone, supplies, and often a personal computer or other tools. If an employee arrives and has nowhere to work, then training and productivity are delayed. If there is an office, but no computer or telephone, productivity is delayed. In either any event, the company wants the asset to be as productive as possible as soon as possible.

Further, timely and appropriate training activities must be scheduled in order to get the employee "installed and started up." These activities may take days, months, or even years, depending on the position and the company. During this time, the employee is typically minimally productive, while collecting full salary and benefits. In the same way that proper installation of capital equipment and effective training of the workforce on its use are essential to the startup phase and to the planned return on a company's investment, identifying, selecting, recruiting, and training the best qualified individual are also vital to making his startup phase effective, so the company may realize a return on its investment in that asset.

A company's balance sheet includes its assets and its liabilities; those assets typically are current assets, property, plant and equipment, and other assets. Unfortunately, nowhere does the balance sheet state the value of the company's human capital assets. It should, or at the very least it should allow, an estimation and notation of that asset value. Most companies, while acknowledging the contributions of its employees, do not think of the acquisition (or disposal) of human capital assets in the same way or with the same thoughtful planning or strategic thinking as they do fixed capital assets.

For example, Table 3-4, which parallels Table 3-3, assigns certain values to each step of acquiring human capital.

So, in this very simple analysis, we have an initial acquisition cost of some $31K to $51K. Since it is a rough estimate to illustrate a point, let's assume an average value of $40K. Suppose Beta carried this value for each employee on its books as an asset that is amortized over time. Would senior managers be more or less inclined to reducing costs through layoffs with these additional charges? Note that *if* that were the case, Beta would also need a policy for treating this cost for employees leaving voluntarily or being terminated, and that event would probably trigger an expense of any balance remaining on the books.

Perhaps more importantly, human assets *appreciate* in value over time through learning, experience, creativity, and so on, making them more valuable, and of course generally receiving pay raises and making a higher salary over time in recognition of that increased value. Let's assume, again for the sake of argument, that this appreciation is equal to their salary increases, discounted for inflation, plus the time and expense spent on training. If we assume the salary increase over 10 years is some 5% per year on a base of $50,000, and inflation is 3% per year, the appreciation of the value of this employee is about $15,000. Let's further suppose we spend some $2,000 per year on the employee developing new skills via training. The total appreciation of the employee would then be about $35,000 over 10 years. Adding this to the initial acquisition of the human asset of, e.g., $40K, gives a total equivalent capital value after 10 years of $75K.

This is a very simplified analysis that doesn't consider other influential variables. And, any new accounting treatment is likely to be more sophisticated. It should, however, be enough to illustrate the point that Beta has considerable money "invested" in its human assets, and that it should at least attempt to value those assets, because as the baby boomers leave, it will be eliminating those assets, some of which may need to be replaced. That expense should at least be recognized from a managerial sense in our decision-making process. Beta will take this into consideration as it moves forward with its plans to add and dispose of human/intellectual capital.

Other Human Asset Valuation Methods

Other, perhaps better, valuation models for intellectual capital are suggested by Thomas Stewart and summarized as follows[10]:

> Market Capitalization to Book Value Ratio. This ratio relates to the difference between market capitalization (share price × number of shares) and the current book value of the company. Allocating this difference in value based on the relative merit of each employee group yields a nominal estimation of the value of various human assets.

Tobin's q. This ratio was developed by Nobel Laureate economist James Tobin, and compares the current market value of an asset to its replacement cost. If the ratio is >1, the asset is worth more than its replacement cost, and the company is receiving a return on its investment. If the ratio is <1, it is not.

Calculated Intangible Value. This term was originally developed by NCI Research, which is affiliated with the Kellogg School of Business at Northwestern University, and it relates to a company's ability to outperform an average competitor that has similar tangible assets. Stewart offers an example calculation of this.

Innovation Value. This is a simple, at least in concept, calculation of the sales or gross profits attributable to new products and services. That value being argued as the human capital's output in innovation.

Knowledge Bank. This model involves "banking" investments made in the development of individuals, e.g., "deferring" a portion of the salaries of new hires since it is in effect being "banked" for development of future revenue. It also adds in R&D expenses as part of the calculated deferral. And, it "expenses," for purposes of the calculation, capital expenditures. The knowledge bank then relates to the deferred value of existing employees as measured by R&D and deferred income.

Employee Attitudes (via surveys), Tenure, Turnover, Experience, and Learning. These measures, although useful from a management standpoint, are even more subjective than those described in the previous text.

The reader is referred to Stewart's book for additional detail regarding all these methods and potential application.

And, there could be other models or variations to these models. The point here is that employees have value, not just cost, and we must recognize both their value and their cost when making decisions, much the same as we do with fixed assets.

But if all else fails, and if Beta finds itself in the position of needing to reduce cost through layoffs, how does it make the right decision? If the above numbers are correct, for every person eliminated, Beta is essentially writing off some $75K in valuable human capital. If Beta is planning to grow the business (and it is, so it can't cost cut its way to prosperity in the long term), and if Beta expects to need those people within a year, then it's likely to be worse off in letting them go than in keeping them. This logic is much the same as that of disposing of a fixed asset. Other questions that Beta should ask include:

1. What is my strategic plan relative to markets and products for the next two years?

2. Do the current plant, process, and equipment meet our needs over that time?

3. How many and what kinds of employees will we need to meet our strategic plan, and are they currently employed?

4. If so, should we keep them (preserving the business and market knowledge, employee potential, training investment, and future productivity), and deploy them to that goal in the interim?

5. Do we have employees who are valuable for their knowledge of our products and/or markets, experience, and other abilities in whom we have already invested, but who can be retrained for areas of anticipated need? Do they have certain tacit or explicit knowledge that we must retain?

6. What will be the long-term effect on customer service and customer perception of the company by our actions?

7. What are some of the core competencies or key value adding functions that we must have strategically?

8. Do we have certain functions that while not strictly "value adding," they are a condition of being in business (e.g., accounting)? How do we manage these?

9. Are we retaining or allowing enough improvement or creative time in our plants?

Finally, Table 3-5 provides several criteria for determining which assets you may want to consider eliminating.

Stephen Covey said "... people have lost their voice because of this industrial age model that no longer works, that calls people an 'expense' and controls them rather than allowing them to unleash their potential."[12] True leadership assures that people do not lose their way, that they are truly considered assets, not just expenses, and that we create an environment where their potential is unleashed. Leadership will be discussed in the next chapter.

Table 3-5 *Considerations for Making Asset Disposal Decisions*

Disposing of a *Capital Asset*	Disposing of a *Human Capital Asset*
The company typically considers:	The company typically considers:
1. Prospective asset revenue	Prospective asset revenue
2. Prospective asset operating cost	Prospective payroll and fringe benefits
3. Strategic business prospects/ margins	Strategic business prospects/margins
4. Benchmark/typical operating performance	Benchmark/typical operating performance
5. Potential for process improvement	Potential for productivity improvement
6. One-time charges	One-time charges
7. Alternative technologies and costs	Contracting for improved "efficiencies"
8. Intangibles	Intangibles
9. Other business/contractual issues	Other business/contractual issues

Summary

Innovation is a key element in a company's survival and prosperity, and the heart of any company's innovational capability is its employees. Innovation must be pervasive throughout the organization, from the shop floor and its "little innovation" and continuous improvement, to the executive suite, with shop floor improvement effectively financing the R&D that aligns processes and products to markets and customer demand. Those employees must be truly viewed as the company's most important asset (not just its biggest controllable cost). Employees must have confidence that their employer, and particularly the CEO, is acting in their personal interests and those interests must be aligned to corporate interests. Outsourcing and offshoring are indeed a constant threat in some circumstances, or could be an opportunity to enter new markets for Beta's long-term success. Beta must convey to its employees the business case for outsource/ offshoring, or not, and then support the employees in their efforts to be innovative and improve their productivity, assuring a strong competitive position in those markets.

The phrase "Employees are our most important asset" must have real meaning, not just be a patronizing statement made at politically opportune times. Valuing those assets using the appropriate valuation model is a good start in the thinking process, and there are many models available that will allow Beta to determine the value of its people, their intellectual capital. This valuation should be done and incorporated and used in the decision-making process, especially as it relates to layoffs. Employees are assets that *appreciate* over time, but are *expensed* on the corporation's financial statements. Capital equipment *depreciates* over time, but is carried as an *asset* on the financials. Accounting's treatment and management's thinking of these issues needs to be updated.

Bob Neurath believes that employees should be treated as strategic assets in whom the company has an investment, not just as overhead to be reduced in the short term without thoughtful analysis. In a downturn, would a company consider all its capital equipment and decide which to sell or eliminate? What company would liquidate its equipment, with the idea it could simply repurchase (think rehire and retrain) new equipment (think new employees) at the appropriate time, almost assuredly at a greater cost? Or might the company evaluate all possible uses for machinery, knowing the strategic plan calls for a return to production in the future? A downsizing that has not been considered from a longer-term view means that much human or intellectual capital, both explicit (about job specifics and tacit knowledge) and implicit (about the company and its products, markets, and customers) has left the company.

There is abundant data suggesting that cost cutting is a high-risk activity and at its best is a short-term approach. Beta understands that innovation and process improvement at every level through its employees is a better approach for business success. Bob Neurath will be asking, how do I get my employees to look forward to coming into work everyday? How do we assure that Beta is innovative from the CEO to the shop floor, and sustain that sense of innovation and creativity? He will strive to provide work that's challenging and stimulating, and demanding and rewarding, providing them with an overarching strategy and the tools to support that strategy, aligned to having Beta achieve exceptional performance. And, he'll work hard to avoid threatening them with layoffs.

References

1. Imai, M., *Gemba Kaizen*, New York, NY: McGraw-Hill, 1997.

2. Moore, R., *Making Common Sense Common Practice: Models for Manufacturing Excellence*, Boston, MA and London, England: Elsevier Butterworth-Heinemann, 2004.

3. Livingston, S., *Pygmalion in Management*, Harvard Business Review, 1969.

4. Begley, S., Editorial on Robert Rosenthal's work, *Wall Street Journal*, November 7, 2003.

5. Schein, E., *Organizational Psychology*, Englewood Cliffs, NJ: Prentice-Hall, 1980.

6. Rath, T. and Clifton, D., *How Full Is Your Bucket?*, New York, NY: Gallup Press, 2004.

7. Csikszentmihalyi, M., *Good Business—Leadership, Flow, and the Making of Meaning*, London, England: Penguin Books, 2003.

8. Burns, David, Personal Correspondence and Discussions, Melbourne Australia, 2004.

9. Liker, J., *The Toyota Way*, New York, NY: McGraw-Hill, 2004.

10. Ford, H., *Today and Tomorrow*, New York, NY: Productivity Press, 2003; original by Doubleday, Page & Company, 1926.

11. Stewart, T., *Intellectual Capital—The New Wealth of Organizations*, New York, NY: Bantam Doubleday Dell Publishing Group, 1999.

12. Covey, S., "Habit Forming," *The Australian Way, Quantas Magazine*, March, 2006.

Leadership and Teams

4

Lead the people; manage the processes.

Hugh Blackwood

Introduction

Leadership is an essential and even dominant element in any company's business success. Data from benchmarking studies and subsequent analysis through statistical correlations indicate "leadership" to be the primary influence on the success of a company in applying best practice.[1] But what is leadership? How do we become leaders? Are we born with leadership qualities? Can we learn it by attending "leadership school," or by using leadership mentors, or should we read the latest books on leadership? As we all know, leadership is simultaneously simple and complex, and it's likely to be context-dependent—what works in one situation may not work in another. It's somewhat like the description of pornography that Supreme Court Justice Potter Stewart once offered: "It's hard to describe, but I know it when I see it." Below are several leadership models offered with a view to help improve your business performance through better leadership, and to identify what these models have in common.

Leaders vs. Managers

Warren Bennis suggests that the following characteristics of leaders and managers[2]:

Leaders:	Managers:
Challenge status quo	Accept status quo
Trust	Control
Innovate and develop	Administer and maintain
Ask what and why	Ask how and when
Do the right things	Do things right
Watch the horizon	Watch the bottom line

Using this model, could we do without either leaders or managers? Most people would agree that we need a good mix of both. Do managers only manage, or leaders only lead? Most would also agree that there are times when leaders manage and managers lead and we're likely to see a mix of behaviors in the same individual depending on the situation. Speaking from personal experience, when I first joined a small, fast-growing, technology company as its president, I spent most of my time managing; we didn't have many of the basic systems and processes in place, and using the phrase organizational structure to describe us was an oxymoron. To allow for effective leadership, or perhaps through effective leadership, we had to have some basic systems in place and operating well. After we managed to establish these systems, it allowed more time to exercise what Bennis characterizes as leadership. Does this suggest that I didn't do any of the leadership activities while I was in my management mode, such as "watching the horizon," or "challenging the status quo," or developing "trust"? No, but it does suggest that the differences between the two, and the effort expended in either, is often context-dependent and difficult to distinguish. I would suggest, however, that as one moves up the organizational hierarchy they need to spend increasingly more time applying Bennis' leadership characteristics and less time applying his management characteristics.

Bennis goes on to say that there are three basic rules: (1) do what's right; (2) do the best you can; and (3) treat others like you would like to be treated. It's difficult to argue with these three as good leadership principles supporting the success of any group. The difficult part is trying to define "right" and "best"; those decisions are generally based on our personal value system. Even his last one, a paraphrase of the "golden rule" can get a bit fuzzy depending on how you like to be treated, but it's still a fairly good rule, and provides a key element of leadership.

Finally, Bennis' definitions provide some insight into managing change. That is, leaders are likely to be better at facilitating change, but once the change has been reasonably accomplished, it must be sustained. Managers are likely to be better at sustaining change. If you're engaged in substantive change in your organization, you may want to consider these principles, either individually, or when selecting people to lead or manage change. We'll discuss managing change more in the next chapter.

Vision, Reality, Courage, and Ethics

In *Leadership: The Inner Side of Greatness,*[3] Peter Koestenbaum describes leadership as being the four points of a "diamond," each point being characterized by the concepts of Vision, Reality, Courage, and Ethics. He suggests that leaders must have *vision* to know where they want to "lead" the company; this vision must be compared with *reality* in order to assess the gap between the vision and the reality. Then they must develop a plan for closing those gaps; the leader must have the *courage* to make the changes needed to close those gaps; risk is inherent in any change process, and there will always be those who are opposed to the change since they have "prospered in the old order of things." All this must be based on the *ethical* treatment of people. Ethics in this instance is about being honest with people and developing mutual trust, treating people with appreciation, dignity, and respect.

Personal Humility and Professional Resolve

In *Good to Great,*[4] Jim Collins characterizes various "levels" of leadership and offers the leadership and management model outlined next, which was developed in a review of the performance of several hundred companies over a period of 30 years, and characterized the leadership of the most successful of these companies:

> Level 5 Executive Leader—builds enduring greatness through a paradoxical combination of personal humility and fierce professional resolve.

Level 4 Effective Leader—Catalyzes commitment to and vigorous pursuit of a clear and compelling vision; stimulates the group to high performance standards.

Level 3 Competent Manager—Organizes people and resources toward the effective and efficient pursuit of predetermined objectives.

Level 2 Contributing Team Member—Contributes to the achievement of group objectives; works effectively with others in a group setting.

Level 1 Highly Capable Individual—Makes productive contributions through talent, knowledge, skills, and good work habits.

He proceeds to discuss certain characteristics of Level 5 Leaders/ Companies, stating that the very best leaders and companies attend to the **people first** and the **strategy second**; recognize that good-to-great transformations take time; and understand three intersecting circles: 1) the area that the company can be the best in the world; 2) the business economics around what they can be best at, and what works best in the business; and 3) what best ignites the passion of its people. Finally, he also indicates they have very disciplined cultures: disciplined people—require less hierarchy; disciplined thought—requires less bureaucracy; and disciplined action—requires fewer controls.

Building Character Through Principles

In *The West Point Way of Leadership*,[5] Larry Donnithorne identifies the key components of leadership as being *courage, integrity, determination,* and *self-discipline. Courage,* for example, would involve choosing the harder right instead of the easier wrong, and putting the needs of the organization ahead of personal desires. Note that this is akin to having a strong sense of fiduciary duty. He characterizes *integrity* as living by one's word, and building character through principled leadership, that is, basic principles and values take precedent over personal desires and consequences. This helps build trust and respect throughout the organization. *Determination,* of course, helps one to work hard and stay focused, to persevere despite potentially overwhelming circumstances. *Self-discipline* is a hallmark of the West Point model; it assures high standards and rigor in all things physical, intellectual, and emotional.

Having been through a West Point experience, I would like to add something that has had a key influence in my career. A concept that was stressed was that the unit commander is responsible for everything his unit does or fails to do. My first reaction to this was that it was unfair. How could I control the behavior of 10, or 100, or 1,000, or 10,000 individuals. What if some "nut case" went berserk and destroyed my unit's performance? Was I responsible for this? With time, however, I came to realize that despite these misgivings about this principle, it forced me to be proactive in anticipating problems, looking ahead for potential failures in performance, and creating an environment that engaged the people in doing the things that needed to be done, and, perhaps more importantly, looking for ways to improve the organization and anticipate potential problems. No, I couldn't control everything in the absolute sense, but I could leverage the knowledge of the people to minimize the possibility of failure and maximize the opportunity for success. And, it forced me to take responsibility, irrespective of the circumstances.

Leadership: Aligning the Organization

In *The Ascendant Organisation*,[6] Peter Wickens states that the greatest risk for an effective leader is to be surrounded by mediocrity and the greatest responsibility is to appoint good people; leaders must have deep personal ambition, an ability to think ahead, and a drive to ensure that their ambitions, both personal and business, are realized. It is paramount for leaders to align the organization so that all are working together to achieve the same objectives.

He believes a leader's personal attributes include:

Intelligence and knowledge

Empathy with people at all levels

Focus on critical aspects for the success of the business

Analyzing logically, thinking rationally, acting intuitively

Uncompromising integrity, honesty, and ethics

Having the wisdom to know when to do nothing

Passion/energy, self-motivation, determination, courage, enthusiasm

Constantly challenging the status quo

He goes on to state that poor leaders (non-leaders?) blame others for their failures, that is, they're not leaders, but bosses, they're not transformers, but conformers. In comparing managers to leaders, he suggests that managers are (perhaps unnecessarily) focused on the present, not the future; are inward looking not outward looking; and focus on brevity, variety, discontinuity, and action, and dislike reflective activities. These comparisons are similar to those offered by Bennis, as previously stated.

Five Practices of Exemplary Leadership

In *The Leadership Challenge,*[7] Kouzes and Posner identify five practices of exemplary leadership:

Model the way (with your behavior, that is, set the example)

Inspire a shared vision (something that is forged, not forced)

Challenge the process (or status quo)

Enable others to act (and experience a sense of ownership)

Encourage the heart (give people a sense of meaning, hopefully aligning personal and corporate interests)

The 21 Irrefutable Laws Of Leadership

In *The 21 Irrefutable Laws of Leadership,*[8] John Maxwell offers 21 "laws" for leadership. Among these are: (1) buy-in—people buy into the leader, then the vision; (2) trust—the foundation of leadership; (3) empowerment—secure leaders give power to others; (4) respect—people naturally follow leaders stronger than themselves; and (5) leadership—determining a person's level of effectiveness.

Inspiring Ordinary People to Extraordinary Performance

Finally, my personal definition of leadership is *the ability to inspire ordinary people to consistently perform at an extraordinary level.*

This is not to suggest anything negative in the use of the word ordinary, but rather the opposite. Ordinary in this context is more in line with the principles illustrated in the book *Citizen Soldier,*[9] where ordinary citizens became soldiers during World War II, and defeated professional armies. These citizen soldiers won because they were inspired by a higher sense of purpose related to freedom and our American way of life. Many sacrificed everything, much the same as our founding fathers.

Within organizations, I think it is appropriate to rely on these same principles, that is, that people want to feel they are a part of something greater than themselves, that they are contributing to superordinate or lofty goals. For example, this goal might be becoming the best company in the world in their market, or it might be involved in saving lives on a daily basis, or it might be routinely giving your customers peace of mind. These should be lofty, perhaps even altruistic, goals that make people proud to be part of the organization. Of course money is important, and of course status is important, but if we can align the organization to a greater purpose we are more likely to achieve exceptional performance.

This concept also fits with Maslow's hierarchy; at the lowest level we must meet our physiological needs, for example having a reasonably comfortable workplace. Next comes our safety needs; for example, working in a non-threatening environment with minimal personal safety hazards. Next comes social needs or belonging to a group. Next comes our self-esteem needs, which are related to having self-worth and a sense of pride. Finally, comes self-actualization, having a greater sense of purpose in our efforts and achievements. Most organizations have achieved the first three: they have a reasonably comfortable and safe workplace that serves as a social entity. Substantially fewer have achieved the fourth goal of giving everyone a sense of self-esteem, and very few have an organization with a high percentage of their people who are self-actualized. That is, they see a greater sense of purpose in their efforts that go well beyond the day-to-day tasks. Moreover, once you've achieved a reasonable sense of satisfaction in the so-called social needs area, money becomes much less a motivator, given pay is perceived as internally equitable and externally competitive. At this point pay is not a motivator in terms of improved productivity, but it can be a substantial demotivator if the pay is perceived as unfair. What motivates people? Creating an environment where most people feel trusted to get a job done, like the work they do, feel in control of that work, take pride in that work (self-esteem), and feel a part of something greater than themselves (self-actualization).

Leaders set the example and create an environment of pride, enjoyment, and trust. Leadership requires a combination of high standards and expectations, and a willingness to let people try, and sometimes fail, and learn, and try again. It is context-dependent. There are times when it would be easy to step in and tell people what to do, but they would not learn. I used to tell people that it was acceptable to make mistakes, but they had the responsibility to correct and make up for them. I also told them that any big mistakes were my responsibility, and that I had to make sure our systems prevented those, and if I didn't it was my fault and not theirs. I believed, and said so, that they were closer to the problems than anyone, and they should take the initiative and act in good conscience and solve them. It freed them to perform at their highest level.

Finally, I think leadership begs the question "how do I get people to genuinely look forward to coming to work every day?" You give them the freedom to be successful for the company, and to give them work that is challenging, rewarding, and calls them to a higher purpose.

Summary: Common Traits of Leadership

Most of these models, if not all, have some common elements:

> Leadership requires vision, a greater sense of purpose, and watching the horizon, albeit grounded in reality.
>
> Leaders put people first, treating them with dignity, respect, and appreciation.
>
> Leaders are trustworthy, true to their word and principles, and create a sense of trust.
>
> Leaders have a passion for excellence, set high work and ethical standards, and create a caring, disciplined, proud environment.
>
> Leaders set the example, and have the courage to support and defend their basic values and principles.

Leadership may be in the eyes of the beholder and hard to describe, but it is easy to recognize. Leaders create an environment where others want to follow them. The examples should provide appropriate food for thought in understanding and applying some basic principles of leadership, and help in your efforts to lead within your company, no matter what your level within the organization, or the heights to which you aspire.

Let's close this section with an observation from Joren van der Veer, CEO of Royal Dutch/Shell Group[10]:

> "The one common value that most leaders lack today, whether in business, politics, or religion, is humility."

Jim Collins also observed that the best leaders "build enduring greatness through a paradoxical combination of personal humility and fierce professional resolve." Most companies, Beta included, will be well served if their leaders take these observations to heart.

Teams and Teamwork

> The best way to get teamwork is to give the team work (to do).
> James Belasco and Ralph Stayer in *The Flight of the Buffalo*

Teams and teamwork are ubiquitous terms in industry today. We hear these phrases repeatedly. But just what does it take to have a good team? Are they effective? Are there times when we should *not* use teams? How do we measure their success?

A simple way of thinking about getting teams to work well together and to have a sense of empowerment is to make sure they have the following traits or characteristics:

- Clear purpose and direction—having a sense of meaning aligned to corporate vision/direction
- Boundaries for the empowerment given, and self determination within those boundaries
- Openness and teamwork being expected
- Skills to operate within the boundaries, or training needs thereto
- Discipline and measurement of effectiveness related to the impact on the business

- Continuing feedback and support
- Flexibility to address changing needs (e.g., boundaries, training, measures, and so on)

While this simple set of principles may suffice to capture key elements, there appears to be more to it than this.

In *Leading Teams*,[11] J. Richard Hackman provides an excellent set of guidelines for creating and managing effective teams based on extensive research. Surprisingly, his research found that when you compare the performance of teams with what is produced by an equivalent number of individuals who work individually, the individuals almost always outperform the teams. He goes on to advise that the reason for the reduced performance is not that teams cannot, or even should not, outperform individuals, but rather, that most teams perform poorly because their leaders have focused on the wrong things in designing or supporting them.

In his model, the five conditions necessary for fostering team effectiveness are:

1. Having a real team, not simply in name only.
2. Having a compelling direction for its work that is clear, challenging, and consequential.

   But, he goes on to state that absolute clarity is not a requirement, suggesting that good direction statements have both specificity and a "little fuzz" around the edges. This "fuzz" is *surplus meaning*. Surplus meaning helps energize the teams who might otherwise fall victim to poor behavior resulting from highly specific goals that have substantial rewards attached. He offers as an example of a good direction statement: Simultaneously optimize safety, efficiency, and customer service.

   He observes that the lack of a clear sense of direction will deprive members of the clarity needed to manage themselves efficiently; and that collective incapacitation results when well-intended exchanges among well-meaning people serve mainly to highlight things about which they disagree.

3. Having an enabling structure that facilitates (not impedes) teamwork

4. Operating within a supportive organizational context

5. Having ample expert coaching in teamwork

When these conditions are present, teams will typically outperform individuals.

He also dispels four myths about teams:

1. Teams whose members work together harmoniously perform better than those that have lots of conflict about how best to perform the work. Not necessarily true: conflict can be a learning experience as long as it is effectively managed. Harmonious teams may not adequately challenge each other or the status quo.

2. A primary "cause" of team dynamics is the behavioral style of the team leader, especially the degree to which he or she is authoritative versus democratic. Not necessarily true; team dynamics is much more complex than this, and a range of team leader styles will work if the five conditions of team effectiveness are present.

3. Larger teams perform better than smaller teams because they have more and more diverse member resources on which to draw in carrying out the work. Not necessarily true: the productivity of a team increases with numbers to a maximum of about eight. His research indicates that "process losses" increase with numbers. Larger teams are more complex and difficult to manage and facilitate communication and decision making. Based on this, he suggests the optimal size for most teams is four to six.

4. The performance of teams whose membership stays intact for a long time gradually deteriorates because members get careless, are insufficiently attentive to the environmental changes, and are too forgiving of one another's mistakes and oversights. Not necessarily true: while this may be true for R&D teams, it is typically not true for normal improvement teams who generally improve their ability to work more effectively over time, and build shared commitment, collective skills, and task-appropriate coordination strategies.

His three criteria of team effectiveness are:

1. The team's product is acceptable to its clients.
2. The team grows more capable over time.
3. The team member's continue to learn, and the experience is satisfying and meaningful. He notes that maximum motivation occurs when there is a 50/50 chance of success; the teams' hurdles must not be so high as to be unachievable, but not so low as to be easy and uninteresting.

He characterizes the essential features of a real team as follows:

1. The task is appropriate for a team. That is, members must work *inter*dependently to achieve an identifiable collective outcome. If that cannot be done (and many times it cannot), then the wise choice is to design and manage the work for individual performers rather than for an interacting work team. A "co-acting group" is not a team; just because members usually work in proximity to one another and have the same supervisor does not mean they are a team. For example, call centers are not considered to be a team in his model. Each member has an individual job to do, and that job's completion does not depend on what the others do.

    Managerial and professional work often does not lend itself to the formation of single teams whose members work only on those teams for extended periods. Instead, one person is likely to serve simultaneously on a variety of different teams that form, reform, and disappear.
2. The task(s) is clear.
3. The team has clear, but moderately permeable, membership boundaries. Having clear membership boundaries does not mean that members must do all their work in the same place at the same time, nor does it mean that membership cannot change as circumstances change. It merely means that members know who is actually on the team and each person's relative role, a seemingly simple matter but one that trips up a surpris-

ing number of teams. He notes that some people are just not good at being on a team, and team membership should be managed accordingly.

4. The team has the authority to manage its work processes. The extent of a team's authority is determined by its ability to set the direction for the team such as collective objectives and aspirations, structuring tasks, established norms of conduct, and available resources. The team also has, perhaps obvious but sometimes not, authority to execute the work, and to monitor and manage the work process, measuring and correcting as needed.

5. The team membership is stable over a reasonable period of time, that time being consistent with the scope of the task.

His research indicates that unclear team boundaries and instability of membership is a pervasive problem for many different types of organizational work teams.

Self-Managed Teams: Caution

He is cautious about the use of self-managed teams, saying that effective team self-management is impossible unless someone in authority sets the direction for the team's work, and he suggests that self-managed teams do not require consensus in decision making. It is usually a good idea to have one person identified as the "leader" of even a self-managed team to facilitate communication and coordination among members. Those in authority can consult widely with team members and other constituents, and increase the chance of getting the direction right, and foster acceptance by members in their decisions. But at some point those in authority must step up to their responsibility and make decisions. And, finally, management must be unapologetic about specifying the ends, while letting the team specify the means to the end under its structure and purpose. Teams are not a democracy. Indeed, democratic ideas and egalitarianism can destroy good teams.

He characterizes the "enabling structure" for effective teamwork as follows:

1. Design of the work the team performs: the work is meaningful; each member feels personally responsible for work outcomes and each member receives feedback on the results of their efforts.

2. Core norms of conduct guide or constrain team behavior. The more members that agree about what is approved and disapproved behavior, the greater the compliance with team norms. An example of norms that work are:

   a. Members should take an active, not reactive, stance toward the teams' environment.

   b. Behavioral boundaries should be demarcated in identifying the few things that members must always or never do.

   c. Teams should not react to whatever comes its way, nor should it always seek harmony.

   d. Norms can be brought in by individual members or others, evolved gradually by the team, or deliberately created as part of the group structure. Norms that are brought in or evolve from the team generally focus on assuring harmony, more so than team effectiveness. Hence, the team is more effective when these norms are created as part of the team's structure.

3. Composition of the team. The three biggest mistakes made when creating a team are:

   a. Assuming that more is better.

   b. Assuming that people who are similar will get along better and create a homogenous team.

   c. Assuming that everyone knows how to work in a group.

The mistake generally *not* made is to overlook members' task-related skill.

An enabling structure generally requires that we give the team the responsibility of assigning subtasks, of coordinating their own efforts, and of switching subtasks as needed so that all the work gets done well while still managing any tendency for social loafing. Autonomy gives teams room to excel, but autonomous teams that do not succeed are indeed very bad.

Organizational Context

If a well-designed team is a seedling, then he characterizes the organizational context as the soil in which it is planted. Three organizational systems have particularly high leverage in supporting teamwork: the reward system, the information system, and the educational system.

1. The reward system: it should provide recognition and reinforcement contingent on excellent team performance. Rewards must be something that team members themselves view as favorable, and at some point such rewards must include financial incentives. It is important to include both intrinsic (e.g., sense of satisfaction in a job well done) and extrinsic (e.g., cash or other financial award) incentives in structuring the reward system. It may also require a multi-tiered approach (e.g., skill development by individuals, team performance, and organizational performance). (My comment: be careful here regarding using a multi-tiered award system. It can be complex and, thus, can be counterproductive and may incite jealousy among team members.)

2. The information system: it must provide the team with data and projections in a timely way. Pitfalls of the information system that should be addressed include:

   a. The good stuff is a secret. Ask yourself which is worse: that the competition might know, or that our own people might not know.

   b. Providers and users speak different languages.

   c. A flood of information is as bad as a drought.

   d. Information really is power, and some hold power unto their own. A better model is that information is the basis for a good relationship; we must share the information that makes us better as a team.

3. The education system: it should make training available to teams for any aspect of the work in which the members are not already sufficiently skilled.

His research indicates that optimal team size for most applications is four to six. He discourages going above eight members. Adding more members beyond this range does not increase productivity because of "process losses."

Expert Team Coaching

Hackman characterizes coaching as being about building teamwork, not about doing the team's work. Three aspects of group interaction have special leverage in shaping team effectiveness:

1. *Effort* applied to their collective work
2. Performance *strategy* appropriateness relative to the task and situation
3. Knowledge and *skill* of the team members

Coaches intervene to minimize process losses or to foster gains in these three areas. Coaching that addresses:

1. Effort is motivational in character.
2. Performance strategy is consultative in character.
3. Skill is educational in character.

The best time to intervene with each of these types of interaction is:

Beginning—Motivational

Midpoint—Consultative

End—Educational

Good coaches focus on the team's task and performance, not on the interpersonal relationships. Emphasis on harmony is often misplaced. How a group is performing shapes the character of the members' interaction, rather than vice versa.

Note: his research indicates that teams that "plunge right in" did better than those who discussed strategy first. His conclusion: it's best to have people work at the task to better understand it before they can effectively discuss how to best go about the work.

Team Leader Execution Skills

He advises that team leaders need the following skills:

Envisioning Skill—Envisions the end state and clearly communicates that to others.

Inventive Skill—Thinks of non-obvious ways to get things done.

Negotiation Skill—Works persistently and constructively with peers and superiors to get resources and assistance needed to support the team.

Decision-Making Skill—Has the ability to choose among various options and uncertainty using perspectives and data.

Teaching Skill—Helps team members learn both experientially and didactically.

Interpersonal Skill—Communicates, listens, confronts, persuades, and generally works constructively with others, especially in high stress situations.

Implementation Skill—Gets things done. At the base level, follows up on the details. At the highest level, manages power and relationships.

His research data indicates that team design is critical, much more so than coaching:

1. Team design is four times as powerful as coaching in affecting a team's level of self-management, and 40 times as powerful in affecting team performance.

2. Good coaching significantly helped well-designed teams exploit their favorable circumstances, but made almost no difference for poorly designed teams. Bad coaching significantly compromised poorly designed teams, but had much less impact on well-designed teams.

3. Teams that are poorly designed and poorly lead are easily out-performed by smoothly functioning traditional units. Teams that are well designed and led easily outperform traditional units.

4. Creative writing should not be assigned to a team.

Corporate Level Observations

Regarding executive leadership, Hackman advises that success-ful organizations almost always are led by a single talented and courageous human being rather than a team. Great leaders cre-ate the conditions that promote team effectiveness any way they can. Effective leaders make sure that the things previously listed are done. This is more important than their personal style. They create real teams, bounded and stable over time, that require members to work interdependently to achieve a common purpose. The team's direction is clear, consequential, and challenging. The team's structure—task, composition, and norms—facilitate good performance processes. The organizational context—reward, information, and education systems—provide the team with the support the work requires. And there is expert coaching for the team to help members minimize inefficiencies and harvest the syn-ergies of teamwork.

The Dysfunctions of Teams

Teams can be dysfunctional. According to Patrick Lencioni in *The Five Dysfunctions of Teams,*[12] dysfunctional behaviors are a result of the following:

Lack of trust

Fear of conflict

Lack of commitment

Lack of accountability

Inattention to results

These seem to be consistent with Hackman's research in that when these conditions are present, a team is less likely to be effective.

Lencioni also offers suggestions on how to avoid these dysfunctions, which are also consistent with Hackman's guidelines.

Summary

Most of Hackman's recommendations appear to be fairly intuitive, but intuition backed by the research he has done provides confidence that these guidelines will provide more effective teams and teamwork within an organization. If we follow these guidelines, and align the organization to a common strategy and purpose, then our teams will be far more effective. People will automatically keep in mind the overarching or superordinate goals and will perform better.

As we noted at the beginning of this chapter, teams must have a clear purpose and direction that is aligned to: corporate vision/direction; boundaries and self determination within those boundaries; openness and teamwork; skills to operate within the boundaries or training needs thereto; discipline and measurement of effectiveness related to the impact on the business; continuing feedback and support; and flexibility to address changing needs (e.g., boundaries, training, measures, and so on).

Beta will apply these principles of leadership and teamwork to the maximum extent possible.

References

1. Moore, R., *Making Common Sense Common Practice: Models for Manufacturing Excellence*, Boston, MA and London, England: Elsevier Butterworth-Heinemann, 2004.

2. Bennis, W., *Managing People is Like Herding Cats*, Provo, UT: Executive Excellence Publishing, 1997.

3. Koestenbaum, P., *Leadership: The Inner Side of Greatness*, San Francisco: Jossey-Bass, 2002.

4. Collins, J., *Good to Great*, New York: Harper Business, 2001.

5. Donnithorne, L., *The West Point Way of Leadership*, New York: Doubleday & Company, 1993.

6. Wickens, P., *The Ascendant Organisation*, London: MacMillan Business Press, 1995.

7. Kouzes, J.M., and Posner, B.Z., *The Leadership Challenge*, San Francisco: John Wiley & Sons, 2003.

8. Maxwell, J., *The 21 Irrefutable Laws of Leadership*, Nashville, TN: Thomas Nelson Publishers, 1998.

9. Ambrose, S.E., *Citizen Soldiers: The US Army from the Normandy Beaches to the Bulge to the Surrender of Germany*, New York: Touchstone, 1997.

10. Huntsman, J., *Winners Never Cheat*, Upper Saddle River, NJ: Pearson Education, Inc., 2005.

11. Hackman, J.R., *Leading Teams*, Boston: Harvard Business School Press, 2002.

12. Lencioni, P., *The Five Dysfunctions of a Team*, San Francisco: Jossey-Bass, 2002.

Note

Material attributed to Hackman is reprinted with permission from Harvard Business School Press and is from Leading Teams by J.R. Hackman, Boston, MA, copyright 2002 by Harvard Business School Publishing Corporation, all rights reserved.

Managing Change[1]

5

There is nothing more difficult to take in hand, more perilous to conduct, and more uncertain of its success than to take the lead in the introduction of a new order of things.

Machiavelli *in The Prince*

Introduction

While Machiavelli is not recommended as a role model for leadership, he did have a fairly good understanding of how difficult it can be to manage significant change. He went on to observe that the innovator has many enemies, those who have done well under the old order; and only lukewarm defenders in those who will do well under the new order. Even those who will do well in the new order recognize the risk involved in change and tend to wait to see how things develop before they commit to the new order. In other words, significant change is risky and requires a compelling reason to support the risk being taken.

Managing change can be extremely difficult. Suppose, for example, you were asked to secure a lasting peace in the Middle East. Managing that change would be an extraordinary accomplishment, since it seems no one has been able to do it for decades, if not centuries. Managing change can also be almost trivial. Suppose, on the other hand, you were asked to help wean your granddaughter, age 3 years, off the bottle and onto drinking from a cup. While this task is not trivial, it could be done in short order if you could suffer through a few tears and a couple days of whining. The kind of change management required in most organizations is, as it is for Beta, difficult, but not insurmountable. It requires *tenacity*.

Let's review the model for managing change that Beta has been using. This is depicted graphically in Figure 5-1. A few years ago, Beta had a compelling reason for change—survival. Now that the company is doing better and is, according to most of the data about average, people seem to have relaxed a bit. Some even appear to be satisfied with current performance. This is not good, and continuing pressures abound in a world where Wal-Mart thinking prevails. Today's mediocrity is tomorrow's failure. What Bob Neurath would like to do is transform the compelling reason for change from "survival" to "becoming the best." So, he's decided to adopt a "Jack Welch approach"—each business unit must be #1 or #2 in its markets, or have a clear plan for achieving that position. And, mediocrity cannot be their standard of excellence. Beta must strive, slowly and quietly, to be the best in the world in each business. No fanfare or banners, just an underlying drive to be the very best.

While no longer steeped in its own history, further change is needed, since Beta still has a long way to go to get to excellence. Getting there will be accomplished with some difficulty. Outlined in Figure 5-1 is

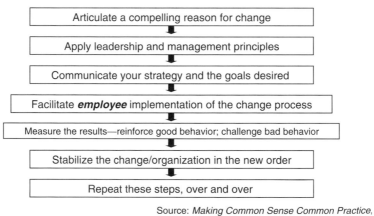

Source: *Making Common Sense Common Practice,*
Elsevier Butterworth-Heinemann Books,
Boston, MA, & London, England

Figure 5-1   *A Process for Managing Change.*

the model Beta is using to manage change. It's not a perfect model, and may not fit all circumstances, but it does address most of the major issues in change management.

These steps are discussed below. Some of these steps will occur in parallel, and some have feedback loops to make changes in the process as learning occurs.

Articulate a Compelling Reason for Change

This is perhaps the most important step in the change process. If there is no compelling reason for change, then change is not very likely. The lack of a compelling reason will allow those who "prospered in the old order" to hold sway over those who are only lukewarm in their support. Change is risky and getting more than lukewarm support requires a compelling reason.

What are some of those reasons? As Beta knows, one of the most obvious ones is that if we don't change, we'll die. That's pretty compelling. Another might be a recent accident or death that was the result of poor practices. Another might be that the CEO has demanded change, or heads will roll. Although this one will have some immediate impact, it's less likely to be sustainable, particularly at the shop floor level, and especially if the CEO is perceived as only "in it for personal gain," and won't be around long. We'll look more at this point later. Most shop floor folks have been around for 20 to 30 years, and have seen it all. Most believe they'll be around when the CEO is gone. In any event, the less compelling the reason for change, the longer the change will take. The bigger the organization, the longer the change will take. That may be just fine. We live in a world where most seek instant gratification—fast food at the restaurant, quarterly profits, and so on. Lasting change takes time, and creating an organization that expects change and continuous improvement takes time. As Jim Collins has observed in his book "Good to Great," transformations take time, as much as 10 to 15 years in the truly great companies. So, if our cash flow will support it, let's take the time to make the changes properly and make them lasting. As noted, Bob Neurath's compelling reason is that each business unit must be, or have a plan to become, # 1 or #2 in its markets.

Apply Leadership and Management Principles

Warren Bennis described the difference between leadership and management as follows[2]:

Leaders:	Managers:
Challenge status quo	Accept the status quo
Trust	Control
Innovate and develop	Administer and maintain
Ask what and why	Ask when and how
Do the right things	Do things right
Watch the horizon	Watch the bottom line

In Bennis' model, leaders foster change and create an environment where change and innovation is the norm, whereas managers stabilize the changes in the organization and assure that they are fully implemented. Which is more important? As Yogi Berra might say: both. As we'll see in Toyota's model for Lean Manufacturing, change and stability are two sides of the same coin. We must have both to do well, and both sets of behaviors are necessary to achieve excellence. However, people who are more prone to have Bennis' leadership characteristics are better at fostering change. Whereas people prone to have manager characteristics are better at stabilizing and institutionalizing the changes once they're made. Does this characterization suggest that individuals will behave exclusively in one mode or the other? Hardly. Beta's experience has been quite the contrary. That is, our behavior is context-dependent, and good managers can move from one mode to another. We must recognize though that different approaches may be needed, depending on where we are in the change process.

Of course, change demands tolerance for risk and uncertainty; when changing a system it gets worse before it gets better, and leaders must create an environment that allows some risk and uncertainty. It also demands strength and determination to sway those who have "prospered under the old order," and coaching and encouraging of those who are "lukewarm" in their support of the new order. Once the changes have been reasonably established, they need to be standardized and managed so that they can be sustained. Once that's done, someone needs to be looking for the next change and/or improvement, and the cycle begins again. As Beta's managers make these kinds of decisions, they'll strive to make sure that the appropriate people are involved in the right leadership and/or management roles.

Communicate the Strategy and Goals

Beta, like all companies, has areas where they are not satisfied with their current performance. That can create a compelling reason for change within a business unit. Once you've identified the reason for change, then a strategy, a set of goals, and the appropriate tools, must be provided and communicated so that people have a clear view of the changes needed, the compelling reasons for the change, the strategy for making the changes, and the ultimate goal or end state. In some cases, Beta is likely to make it clear that not achieving the desired goals is not an option. Exactly how and when they are achieved is an option and is something that requires the support and hard work of everyone. Of course we'll need to be able to measure against those goals to help assure that we've met them, and some suggestions on that are provided below.

A word of caution may be appropriate here. The strategy needs to be reasonably well defined so that it can be understood by the employees and can assure alignment of the organization to the strategy and goals. As we've discussed, one of the most important things a leader can do is to align the organization to its strategy and goals. However, it must not be so detailed that it leaves little room for change as we proceed with its implementation, and each step of the change process has a feedback loop that allows for improving the process itself. In his book *On War*, Von Clausewitz[3] observed that developing a strategy without understanding the tactics and capabilities of the troops was folly. So it is with our strategy and goals. We must fully appreciate the need for tactics and understand the capability of our people, and we must be flexible enough to allow modifications to our change management process as we learn. Our people will implement the strategy. They must have a strong hand in developing the day-to-day tactics for its implementation. And, there must be a feedback loop that allows us to accommodate changes to the strategy based on what we learn as we tactically implement those changes.

In setting goals, one of the more common techniques is to use benchmark data to determine gaps in the current performance and then seek to close the gap by achieving "benchmark" performance. Be careful using benchmarks. Using benchmarks is an excellent way to determine the gaps in your performance as it relates to certain measurements. What it often fails to do is provide an understanding of how that performance was achieved, or a process for closing the gaps. It may be that the gap you're worried about is not one that's appropriate to be closing at this point in the business. An example might make this more clear. One of the more common "big gaps" in benchmark performance in manufacturing plants is maintenance overtime. A common maintenance overtime rate is 10% to 20%, and sometimes more. So-called benchmark performance is in the range of 3% to 5%. So, it's easy enough to achieve benchmark performance on overtime. We order people to minimize or eliminate overtime. Nevertheless, the consequence of that may be that work doesn't get done and products don't get delivered. It's better to ask how we stop the defects that are creating the failures and in turn are creating the need for the overtime. We need to be systems thinkers and understand the consequence of our decisions at a business or operational level as we implement the change process. One of the best sources for understanding the consequences of our decisions is the shop floor. They deal with those consequences every day.

Facilitate Employee Implementation of the Change Process

We've heard many times that people don't want to change. That may or may not be true. What Beta has found is that people will change if: (1) given a compelling reason to do so; and (2) *if they participate in the change process*. Or, as Margaret Wheatley said, "People own what they create." Once a compelling reason has been established, facilitating employee participation in defining the specifics of the changes provides them with a sense of ownership and control in the change process. Having a bit of ownership and control is a lot less stressful too! Sustaining that sense of ownership takes time. Winston Ledet's experience has been that *every employee* must participate in *at least one improvement effort, at least once per year, for at least three years* to assure large-scale, lasting, cultural change. The logistics of actually doing this can be daunting. Deming said it takes seven years. Jim Collins said 10 to 15. The point is that lasting change takes years, and in fact, we must constantly improve just to remain competitive.

Thus, a process is needed by which the ownership of change is facilitated at the shop floor level. This process must include sufficient funding, time, training, and in some cases a facilitating organization that can be internal, external, or both. Beta has often used external resources to initially facilitate organizational change. As Deming said, "Profound knowledge comes from the outside and by invitation. A system cannot know itself." However, it's important that these external resources not be introduced with banners and fanfare. Doing that suggests it's just another passing fad. The shop floor then acts much like a rock at the seashore: they see the "next wave" coming and just sit tight. This too will pass. The change process must have, as Deming also said, *constancy of purpose,* and introducing any new tools needs to be done with clear communication about how it supports the company's overall strategy.

Some might argue that some of the tools are strategies themselves, and they can be, depending on the level in the organization at which they are applied. To determine which is which, first articulate the business strategy and then incorporate the tools that best support the strategy. Otherwise, the tool may drive the strategy instead of the strategy driving the tool.

It's important to celebrate and build upon the early successes in a low key way, to learn from the early failures (the best learning often comes from failure), and to continue to foster and reward a continuous improvement environment. But, the concept of "low hanging fruit" is more often than not largely a myth, since the improvement targets are constantly moving. The low hanging fruit picked today may end up being replaced by other fruit tomorrow, so a process for continuous improvement is essential, always looking for the next opportunity. As in athletics, you have to work really hard just to keep what you have. You have to work even harder to improve.

Measure the Results: Reward Good Behavior, Challenge Bad Behavior

As Joseph Juran said, "if you don't measure it, you can't manage it." On the other hand, don't measure it if you don't intend to use the information. The measure loses credibility. Once the strategy and its related goals are established, a set of measures must be put in place that assures we're achieving the appropriate goals. Start with high level business measures that reflect accomplishment of the goals related to the strategy. Some examples that Beta uses include unit cost of production, return on net assets, gross profit per product line, and so on. These are *lagging* indicators, and a bit like looking in the rear view mirror to see where you've been. As such, these lagging indicators should be "cascaded" down through middle management to the shop floor, where you'll need some *leading* indicators. These are the things that you must do to improve the lagging indicators. Some examples here include process conformance and control, planning and scheduling compliance, housekeeping, average component life, and so on. For example, if I want my unit costs to decrease in order to increase my gross profit and return on net assets, I have to spend less. To spend less, I have to eliminate the defects that are resulting in the failures that induce the extra costs. To do that, I have to improve the consistency of my processes—less variability, higher quality, less waste. To do that, I have to measure and improve my process conformance. Leading indicators measure your ability to do the right things day to day. Lagging indicators measure the *results* of doing the right things day to day. Finally, whether a measure is leading or lagging can depend on where you are in the organization. Leading for one might be lagging for another.

Reward good behavior. Whatever you reward, tolerate, or subsidize, you'll get more of it. The federal government has demonstrated this principle for decades; on a personal level, as parents, we've known for years that our children will find the limits of our tolerance, and then exercise them. For our purpose, however, rewarding good behavior is more appropriate. This is not about pay for performance, since the data generally say that pay for performance does not work very well and its effects are not lasting. Perhaps "reinforce" would be a better word to use than reward, since rewards are usually associated with money. Reinforce those behaviors that have resulted in improvements in the leading or lagging indicators. This includes doing things like thanking people for a job well done, having them do a presentation to the boss, announcing the results of their efforts in a newsletter, asking their opinion on a problem area, facilitating their participation on improvement teams, and so on. Showing your respect and appreciation for a job well done motivates more effectively than money or banners.

Challenge bad behavior. Any time you go through a change process, you can expect some "casualties." Jim Collins believes that we need to get the "right people on the bus" and the "wrong people off the bus." Getting the wrong people off the bus will result in "casualties." If despite your best effort to engage people in the change process you find that some are simply unwilling to make the changes needed, then swift action must be taken to make sure people understand how serious the need for change is. Tolerating foot-dragging, recalcitrant behavior, and passive-aggressive behavior (agreeing but not acting) will scuttle any effort to change the organization. This is particularly true if these behaviors are being exhibited by people in leadership positions. When action is taken in this regard, the people must still be treated with dignity and respect.

Stabilize the Organization in the New Order

Once substantial changes have been achieved, these changes need to be stabilized in the new order of things. Practices and procedures need to be put in place or updated. The appropriate people need to be trained in the new procedures. Our compliance to these must be audited, mostly by highly skilled immediate supervisors. The results we're getting from those practices must be measured, and the next improvement must be found. As mentioned, a new practice may need to be done 21 times before it becomes habit and a part of our new culture and behavior.

Another technique for stabilizing the new practices is to incorporate some of these into our International Standards Organization (ISO) 9002 or other quality certification requirements. The addition of these should be specific enough so that they become part of our standard work, but not specific enough so that opportunity exists for even further improvement without changing the requirements, nor do we get cited for failure to comply because it is too specific in every requirement. Once stabilized and habitual, it's essential to continue to look for additional things to refine and improve our practices. In doing so, a culture that expects and welcomes change will become an ordinary part of the culture. This is perhaps the hardest part of all, creating a culture where change is the norm. In this Utopian world, the organization has the inherent capacity to change, quickly and smoothly, in response to changing organizational goals, customer requirements, and competitor challenges.

Finally, stabilizing the changes in the new order likely requires most organizations, and especially Beta, to begin thinking about the stability of its management teams at each plant, particularly the plant manager. This is discussed below.

Succession Planning and Management

One area that's of particular concern for Beta regarding managing change, or more particularly regarding assuring stability and sustaining the changes in the organization, is that of management tenure. There is an apparent lack of stability in the management team at any given plant. Beta has analyzed its plant managers for "years in role" in each business unit, and found that about one-half have only two to three years as a plant manager at a given plant before moving on to another plant or another job either inside or outside Beta. In some Beta business units, plant managers change every one to two years. This is cause for concern. It is Bob Neurath's view that this results each time in shifting priorities, adjustments to personnel changes, and generally inducing instability throughout the organization. This kind of change is not the kind of change he's seeking. As Hugh Blackwood observed, the employees in this situation often become "webees": we'll be here when you'll be gone, so we'll not be changing much. And, within Beta, it has often happened that a plant will do well under one manager, only to see it decline under another, or vice versa. Most people who have been around longer than a fortnight have seen this happen, some of us many times. How can Beta do a better job at succession management? What are some of the imperatives for succession management? The answers are not always obvious, but below are some experiences at Beta, that will hopefully stimulate debate, and help the business do a better job with succession management.

Beta's Omega Division

Some years ago, Beta's Omega division, a large multi-national, multi-billion dollar subsidiary, had a policy of changing plant managers, and most of the plant management team, every couple years. The ostensible goal behind this was to expose the maximum number of "fast track" managers to the responsibility of running a relatively large organization. At the same time, because of pricing pressures induced by market forces, there was considerable pressure for cost cutting. And, operating and maintenance practices were typically no better than average, and for that matter, not very consistent from plant to plant.

The result of this approach is perhaps obvious. Each new manager sought a short-term reduction in costs through staff and budget reductions, but did not have the patience (or time) to change the underlying practices and processes. Each had to be "successful" in cost reduction in less than two years. Moreover, each manager was gone by the time the consequences of their cost cutting really had its impact on the performance of the plant (e.g., reduced process and equipment reliability, reduced production, lower quality, and higher *unit* costs). Keep in mind that costs are a consequence of your business system design and your practices. The natural result of all this was a death spiral, for each new manager was faced with the same problem of cutting costs to be more competitive, all the while equipment and processes were continuing to decline, and there was no time or money to correct them. Today this division is a wisp of its former self. By nearly any measure, sales volume, gross profit, market share, and so on, it is some 10% of the level it was during its better times. It was one of Beta's divisions that barely survived, and is still on the "strategic watch list." It appears that this approach of changing managers frequently was, among other reasons, a key contributor to this company's demise. You cannot have stable processes when you change the management team or "process" every two years.

As we'll see later, Lean Manufacturing, a methodology that evolved from the Toyota production system, requires that we have process stability as a foundational element. This can be interpreted as it being essential to have reliable equipment and good process control. While these three elements are certainly essential for process stability, they do not address the more fundamental issue of leadership stability, and having the constancy of purpose that has been stressed in doing what is necessary to make a company successful. According to Jeff Liker, Toyota stresses, among other things, the need for stable personnel, slow promotion, and careful succession.[4] It is very difficult to have organizational stability and constancy of purpose when the management structure changes frequently and is highly unstable.

Better Succession Management

How do we achieve stability in the organization through better management selection and succession management? How do we get and keep good managers in our organization? How do we assure that each successive manager will sustain the existing improvements, and make the plant even better?

Before you can solve a problem, it's important to recognize that you have one. Until now, Beta has not recognized that frequently changing managers is in fact a problem, one that induces instability in the organization through shifting management styles and changing priorities. Toyota is generally recognized as one of the best manufacturers in the world. As we've discussed, they believe that slow promotion and *in*frequent changing of managers is a better business approach, keeping its managers in their positions for several years. But, they still expect continuous improvement and better performance in each subsequent year and from each subsequent manager. With this, managers must be both tactical and strategic. Repeating Bill Brenneman's observation, "The capacity to be strategic lies in understanding the consequences and time scale of the actions I take at my level of authority." Toyota expects its managers to improve each day-to-day operation, but never at the expense of long-term business capability. One of the key problems in U.S. manufacturing today is executive's short-term thinking, and the "musical chairs" associated with the management team (the frequent changing of managers that accompanies short-term thinking).

Managers should be in a position long enough to suffer the consequences or reap the rewards of the job they've done. Plant managers, barring extraordinary circumstances, should be in that position for a period of approximately five to seven years. Most "systems" have a lag time between the "stimulus and response." That is, from the time a significant decision is made in a business, it can take months, and more often years, before the full results of that decision are realized. In large businesses this lag time is typically one to five years, depending on the nature of the change. Granted, cost cutting this year may result in an immediate reduction in our costs, but the true impact of that cost cutting would not likely be known for two to four years. The system (people) will try to adapt to the reduced resources, critical actions may be delayed, training may be deferred, and so on, and the consequence of that often takes years to make its true impact known. This view is supported by Gary Hamel who reported that 86% of the companies he studied suffered a major downturn in performance after three years of cost cutting. It's likely that during the first two years the cost cutting seemed to be working, but finally caught up with them with its full consequence in the third year.

Selecting the Right Managers

Most companies need a better management selection process. Having the experience of looking at several manufacturing plants, Bob Neurath is disappointed. Most if not all the employees are good people, but too many are misplaced, or have been given short-term objectives that are inconsistent with Beta's longer-term goals. Getting the balance right is difficult, but Beta needs considerable work here. Perhaps due to enormous Wall Street pressures, most executives are primarily driven by short-term goals. While they may mouth words about being strategic, their behavior belies their words. For whatever reason, many lack the "capacity to understand the consequences and time scale of the actions they make at their level of authority." And, as a shop floor technician once observed, "Some managers are better at managing their careers than they are at managing the business." That is, they're good at political maneuvering. So how do we correct this? While too comprehensive to cover in this section, there are two good books that provide excellent guidance for getting the right managers in the right roles: *Good to Great* by Jim Collins, and *Executive Leadership* by Eliot Jacques and Stephen D. Clement.[5,6]

Several recommendations that managers should follow include:

1. Demand so-called best practice, but be supportive and encouraging by providing the time, tools, training, measures, and rewards to support a best practice standard.
2. Be comfortable going to the shop floor, seeing the work processes for themselves with a critical eye, and talking with people and supporting their ideas for improvement—the people doing the work are most familiar with the work and so must be routinely supported in improving the work processes.

We must remove the obstacles to their success, since *nothing changes until the shop floor does things differently*. We must help all our employees to look forward to coming to work, and make the day-to-day tasks just a little bit better. Toyota refers to this as going to Gemba, or the workplace, and seeing for yourself.

3. Have a good understanding of the technology and processes within the plant, or have a means to achieve that *before* they assume the role of plant manager. While it may be true that the higher in an organization you go the less you need to understand the business itself, it is critical for a plant manager to have a good understanding of the plant processes.

4. Be held accountable for both leading and lagging indicators. If we're doing the right things (leading indicators), we'll get the right results (lagging indicators). At the same time, we must take a strategic view on the lagging indicators, and they must be sustainable. It does little good if we improve our unit costs one year through cost cutting, while the people skills and equipment conditions are allowed to deteriorate to the point of not being able to sustain the required level of production.

5. Go through a battery of tests that characterize certain attributes required for the job in terms of desirable and undesirable traits. Certainly these tests can be "gamed," but having good testing improves the odds of getting the right people in the right roles.

Sustaining Improvement

Let's suppose that we have in fact achieved a higher level of performance, through applying best practice standards and, as determined by measures such as Overall Equipment Effectiveness (OEE), lower unit costs, higher gross profits, and longer equipment life. How do we sustain these improvements? How do we prevent the next manager from performing well in the short term by saving money in not doing the things that got the improvements, to the detriment of the future of the business? Some actions to be taken are discussed above.

Other actions may be less obvious or less common. For example:

6. Making those key actions identified a part of the plant manager's performance objectives. For many plant managers, if it isn't on their objectives, it's not important. One objective we might impose on them is to identify and implement a specific number of improvements in the future.

7. Having the current plant manager *and* department management formally present to the new plant manager their current practices, measures, performance, historic improvements, future plans, and so on, so that the new manager understands their achievements, obstacles, needs, goals, and plans for the future. Of course, the new plant manager should meet with each of their immediate reports and get an informal understanding of how things work.

8. Recognizing, as Toyota has, that innovation and standardization are two sides of the same coin. We must constantly improve our practices on the shop floor or innovate our practices, but once the change is made and standardized, then we must be continuously looking for our next improvement. As Henry Ford said "If you think of 'standardization' as the best that you know today, but which is to be improved tomorrow—you get somewhere. But if you think of standards as confining, then progress stops."

Finally, a few additional suggestions are provided. Winston Ledet advises that sustaining change requires that we make our new procedures and processes a "habitual way." This habitual way must the path of least resistance or the easiest way to get the work done. Put another way, he says we must make it easy to do the right thing, and make it hard to do the wrong thing—so that the easiest way is the right way.[7] Other techniques we might use to foster the "right way" include:

1. Poka Yoke (mistake proofing in the design)
2. Training
3. Mentoring
4. More definitive procedures
5. The use of superordinate goals in driving employee behaviors

Good succession management requires exceptional discipline and a culture that insists that we sustain what we've achieved, but be constantly looking for the next improvement. Beta International is determined to apply these principles in achieving better succession management.

Summary

What does it take to be a world class athlete? A little talent, and a lot of very hard work; you have to be *tenacious*. What does it take to stay in world class condition? A lot more hard work and *tenacity*. Managing change, putting a good succession plan in place, being fair and equitable in the compensation of all employees, and aligning the organization is much the same. It begins with a compelling reason for change, that's clearly communicated to employees along with a strategy and set of goals. This must be followed by a process for facilitating the changes and for creating a sense of ownership in the people making the changes. The results must be measured, and good behavior must be reinforced. Bad behavior must be challenged. Once the change has been reasonably accomplished, it must be standardized and inculcated into the organization. Once this is successfully done, the process must be repeated, over and over, until a culture is created where change and continuous improvement is the norm, and it's mostly owned by the shop floor. Following this process will improve the odds of being a world class performer.

Let's move on to the various strategies and tools and the process for selecting these.

References

1. Moore, R., *Making Common Sense Common Practice: Models for Manufacturing Excellence*, Boston, MA and London, England: Elsevier Butterworth-Heinemann, 2004.

2. Bennis, W., *Managing People is Like Herding Cats*, Provo, UT: Executive Excellence Publishing, 1997.

3. Liker, J., *The Toyota Way*, New York: McGraw-Hill, 2004.

4. Von Clausewitz, C., *On War*, London: Penguin Classic Books, 1982 (original published in 1832).

5. Collins, J., *Good to Great*, New York: Harper Business, 2001.

6. Jaques, E., and Clement, S.D., *Executive Leadership: A Practical Guide to Managing Complexity*, Malden, MA: Blackwell Publishers, 1994.

7. Ledet, W., "Getting to Stage Three," *TMG News*, The Manufacturing Game, October 14, 2005.

Business Level Failure Modes and Effects Analysis: Selecting the Right Improvement Projects and Tools[1]

> If the only tool you've got is a hammer, everything pretty much looks like a nail.
>
> *Bob Ford*

Two of the more common questions asked in Beta's manufacturing plants are: (1) where should we start with our improvement effort? and (2) what improvement tools best apply to our business? Most of Beta's divisions are looking for a relatively easy way to identify the biggest opportunities they have for improvement. And, many managers are trying to become "systems thinkers," that is to think of their manufacturing plant as a complex system that must be fully understood as to inputs, outputs, and variables. This is in contrast to being "event thinkers," wherein one reacts to an event without fully appreciating all the actions and circumstances that led to the event occurring.

Essentially all Beta's plants have some form of site-wide strategy and improvement plan, and as we've seen, the tools they are applying are numerous. They've been bombarded with promotional material and success stories regarding the use of tools and methods such as Lean Manufacturing, Six Sigma, Supply Chain Management, Total Productive Maintenance, Reliability-Centered Maintenance, Root Cause Analysis, and Kaizen, just to name some of the more common ones. They've also used all these in one form or another, and it seems that each one if properly applied, works well, or if not properly applied, does not. The "best" tool seems to depend on the particular business situation, and given that, there may not be one "best" tool. The question that many managers ask is how do I know which one to apply and where to start? Below is a model that Beta has successfully used to help determine where to start and that provides insight into what tools to use.

The Manufacturing Plant as a Business System

First, a manufacturing plant should be thought of as a business system, and management and the shop floor must understand the events or defects (anticipated or otherwise) that occur which may result in the loss of capability of that system. In a perfect world our plant would run perfectly all of the time. We would have perfect quality, maximum production rates, no unplanned downtime, and instantaneous product changeovers. Of course none of us has ever seen that world. With that in mind, we should understand those events or defects that cause us to lose capability, and ask if they are acceptable to the business or not. At times, they will be perfectly acceptable (e.g., rapid product changeovers, capital project upgrades, planned maintenance, and so on). At times, they will not. We must understand the causes of lost capability, and then take appropriate action to minimize them. How so?

One of the more common methods for identifying important problems is to measure Overall Equipment Effectiveness (OEE), and particularly the losses from the ideal that goes with this measure. The losses from the ideal are typically due to planned and unplanned downtime, rate, quality, changeover, and other losses in which some are unique to the business. These are then used to select and apply the appropriate improvement tool to reduce the losses being incurred. However, suppose we're not measuring OEE or its equivalent, what then? Or suppose we don't have confidence in the numbers; they are "guessed at" monthly in a report we don't use, what then? Or suppose we'd like to understand the system-level interactions and be able to make better decisions on that basis. OEE will not provide all of the details of production losses at a system level.

Business-Level Failure Modes and Effects Analysis (FMEA)

Suppose our costs are too high: how do we know where to focus our resources to lower those costs? Or, perhaps that's the problem, we have too many "resources" with their attendant costs, so we cut costs, with the expectation that costs will decline, only to see our production capability suffer. Things are often not what they seem to be. For example, when maintenance costs are too high, companies often cut them, to their detriment. Rather than cut maintenance costs, perhaps we should understand the source of the defects that are resulting in the failures and the subsequent need for maintenance. The model below is one that Beta is using and provides a step-by-step process for understanding where the "defects" are, that is, those things that cause us to lose production capability or incur additional costs, so that we can eliminate the defects from our business system.

The steps are:

1. Create a simple block diagram of the production process or line.

2. Assemble a cross-functional team from each production area, and, as needed, a support function team. The cross-functional team from each area typically consists of a senior operator, an operating supervisor, a senior mechanic, electrician, or technician, and a maintenance supervisor. It may also include others, such as a vendor and/or an engineer for certain critical equipment. The support team is typically composed of a stores supervisor, a utilities supervisor, a human resource and/or training manager, a purchasing supervisor, and a capital projects engineer.

3. The heart of the analysis process is this: a *failure* in the business system (the production line) is defined as *anything* that results in *loss of quality production output* or in *extraordinary costs*.

4. Next, each cross-functional team reviews each step of the production process, identifying failures in the production system at each step of the production process. If there are area teams using this technique, it's usually best if each area does its analysis independently, and then presents its findings to the other teams at the end of the day.

5. The support functions must also be reviewed for failures in our business system (e.g., poor quality or quantity spares, insufficient utilities, and so on).

6. During the review, the value of each business-level production failure is analyzed and estimated/calculated: type of failure, and its frequency per year × losses per failure × value per loss to estimate the dollar value, or quantity of product "lost" per year.

The process is shown pictorially in Figure 6-1.

Each area team asks the questions shown above and depicted in Figure 6-2.

For example, supposed in step one of our production process, we identified supplier quality as a major problem. That is, once a month the supplier provided raw material of substandard quality. This quality problem required adjustment of the process and resulted in lower yields. Suppose we calculated that this lower yield resulted in the loss of the equivalent of 10 tons of product (or 1,000 units or 1,000 gallons or whatever appropriate unit), with each ton having a gross profit of $1,000. This would result in a loss of $120,000 per year.

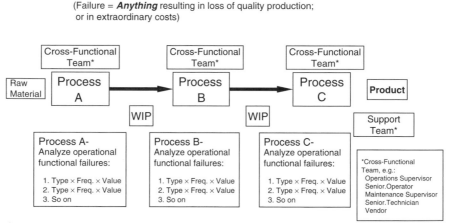

Source: *Making Common Sense Common Practice*, Elsevier Butterworth-Heinemann Books, Boston, MA, & London, England

Figure 6-1 *Model for Performing a Business-Level Failure Modes and Effects Analysis (FMEA).*

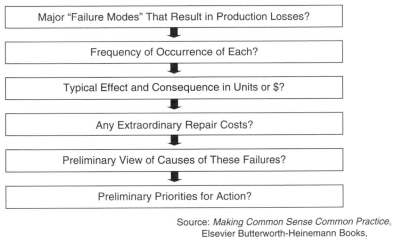

Major "Failure Modes" That Result in Production Losses?

Frequency of Occurrence of Each?

Typical Effect and Consequence in Units or $?

Any Extraordinary Repair Costs?

Preliminary View of Causes of These Failures?

Preliminary Priorities for Action?

Source: *Making Common Sense Common Practice*,
Elsevier Butterworth-Heinemann Books,
Boston, MA, & London, England

Figure 6-2 *Business-Level FMEA Questions.*

Further, suppose other disruptions for accommodating this poor raw material were valued at $2,500 per event. lost productivity of people, plant re-configuration, and so on. Overall, the value of this failure in our system would be estimated at $150,000 per year, a fairly substantial sum. We continue with this process until we've identified all major losses in each production area, as well as those attributable to the support functions.

Next, we prioritize those losses that provide the greatest opportunity for improvement, and select the appropriate tools for eliminating or minimizing them. Note also at this point that we do NOT try to solve any of the problems we identify. We will prioritize the results of our analysis using the model in Figure 6-3. For example, after we analyze our business-level failures and place a nominal value on them, we'll next "plot" them on a chart similar to Figure 6-3. Tasks that are easy to do and have the most value, get done first; tasks that have less value, and are easy to do get done next; tasks that have high value, but are difficult to do, for example, changing a major technical process or getting capital funding, are done next. The last category, tasks that have low value and are difficult to do, may not get done at all, at least not in this round of improvements. For example, at Beta's Orkney plant, task #1 in Figure 6-3 required a major design change to the process, including considerable capital. It provided high value,

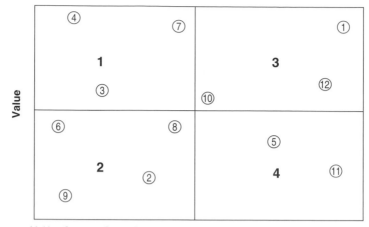

Source: *Making Common Sense Common Practice*, Elsevier Butterworth-Heinemann Books, Boston, MA, & London, England

Figure 6-3 *Decision-Making Model.*

but was also difficult to do. Whereas task #4 provided high value, but was relatively easy to do. This had to do with getting operators trained to follow a particular procedure, one that they had not been doing because of a lack of training and understanding of its importance.

A word of caution in doing the analysis relates to identifying preliminary or potential causes. Very often the team will want to begin solving the problem after they've suggested some potential causes. This should *not* be done at this time. The purpose of this question is to get people thinking about potential causes, but not to solve the problem. We won't begin problem solving until after we've plotted the opportunities in the decision-making model chart. After that we can select those problems that provide the highest return for the least effort given our resources, and we can select the right tool for the problem solving. Too often people want to immediately begin problem solving. Hold that until after the analysis is completed.

It's also important to point out that this process is imprecise and will require validation of initial findings. One of the key benefits, however, is that people are working as a team, using a common strategy that's focused on the success of the production line; it has a business system focus for the teams. And, the team develops an action plan to improve the overall system performance. That in itself is invaluable.

Typical Results

Beta has found the following to be fairly common problems at almost all its plants when doing a business-level FMEA. Case studies will be provided later that illustrate these problems:

- Lack of understanding of the pacing unit for setting production requirements
- Lack of understanding of the impact of upstream and downstream area performance on each other, and the plant's overall performance
- Changeover time and setup/startup problems
- Production stops for breaks and lunch
- Raw material problems: poor quality and/or insufficient quantity
- Erratic production planning, primarily driven by erratic sales forecasts
- Equipment failures
- Spare parts unavailability or poor quality
- Design/layout features making maintenance difficult
- Poor power quality resulting in electronic problems and failures
- Operator inexperience and lack of training resulting in inconsistencies and failures
- Inadequate lubrication resulting in machinery failures; or lubrication delegated to the operators without adequate training, or buy-in
- Mechanics in need of training on critical equipment and/or precision skills

Selecting the Right Tools

Using this approach to better understand and prioritize production losses and extraordinary costs, we can now do a better job selecting the appropriate tool and approach to solve or mitigate the problem. And our strategy will be based on a system-level review of our production process. For example:

1. If supplier quality was a top priority problem (opportunity for improvement), we could apply supply chain principles, and perhaps Six Sigma in an effort to reduce the variability of their supply.

2. If the next biggest opportunity was inconsistency in our production process due to a lack of training in our operators, we would probably ask human resources or training to develop a plan to address this need.

3. If the next biggest opportunity was related to a specific machine and its unreliability, we might apply Reliability-Centered Maintenance (RCM) initially, to better understand the machine's functional requirements and the failure modes that are resulting in loss of functionality. Next, and if appropriate, we might apply certain Total Productive Maintenance (TPM) principles (e.g., operator care, or better preventive and predictive maintenance to better manage these failure modes). If the problem with the machine was particularly difficult, we might use Root Cause Analysis (RCA).

4. If the next biggest opportunity was to reduce the erratic nature of production planning and frequent changeovers, we might work with sales and marketing to analyze our product mix, sales and gross profit by product, and key customers, with a view to better understand and rationalize our product mix and sometimes even our customers. We would probably also want to implement a quick changeover capability, and work to level our production flow, even at the risk of modestly increasing inventory in the short term.

5. If the next biggest opportunity was to reduce spares unavailability, we would probably want to apply RCM methods to understand our most common failure modes, or our highest consequence failure modes, and then use certain supply chain principles to make sure we had critical spares on hand.

Using the business-level FMEA to analyze a particular production system at the system level seems to work well in helping quantify the opportunities and prioritizing them for further action. We need to understand the business consequence of each of the major failures in our business system, and then select the appropriate tool(s) or strategy to address these opportunities. Case studies illustrating the results of using the business-level FMEA are provided next.

Beta's Allen Central Plant

At Beta's Allen Central plant, a business-level FMEA was performed on the entire production line. This analysis indicated, among other systemic issues like training and planning, that improvement in two separate sets of machinery was essential to improving the overall capability of the plant. These two sets were constraints in the production process, and so if they could be dramatically improved, then overall plant performance should also improve dramatically.

For one set of 16 machines, an analysis was done using the RCM methodology for identifying the machine's functional requirements and failure modes, including a criticality analysis of each significant failure mode. An assessment of each machine's condition was also done using operator input, as well as inspections and more sophisticated predictive maintenance tools. The results of this review were used to make decisions about restoring the appropriate machines to "like-new" condition and performance (a TPM principle) in a planned manner.

The process of making all 16 machines highly reliable took some six months. Initially, a team of people was established to "restore" a given machine to like-new condition and commission it to a specific set of standards before returning it to production. In the interim, the other 15 machines were kept operating by another team instructed to "keep 'em running" by whatever means appropriate. Once the first machine was restored and returned to service, the next machine was addressed, typically the machine in the worst condition at the time. This process continued, with routine communication between the two teams, until all 16 machines were restored to like-new condition. After this, it was typical that one machine would be down for restoration, typically the machine in the worst performance condition. All the machines were also routinely monitored as to their condition and any developing functional failures, so that information could be input into a plan for resolution.

In the second set of 20 machines, maintenance would be routinely called in to "fix" a problem, typically a sudden trip of the machine. These call-ins numbered about 70 per week. Typically, after the operator called the maintenance technician, he or she would leave without waiting for the technician. After the fix, the technician would call the operator to return, and leave. Amazingly, the two rarely met to discuss this chronic problem. A quick at-the-scene analysis revealed that in the short-term training the operator to perform a simple "re-boot" of the system would resolve the immediate issue for needing a call-in. This reduced the call-ins from 70 per week to one per week. In the longer term, it was necessary to correct the software and firmware in the machine. And, perhaps more importantly, the Allen Central plant initiated a requirement for the operator to stay with the machine and discuss the problem with the technician so they could both learn from each other. Would you leave your car at the shop with instructions to fix it, leave, and then not ask what was wrong when you picked it up and paid the bill? Wouldn't you have some communication with the repair shop about the nature of the problem, the cost of the fix, and potential actions for avoiding future problems? If so, why don't we do that for the machines that help pay for our car? Clearly we should.

Beta's Swampton Plant

At Beta's Swampton plant, applying the business-level FMEA resulted in finding that each production area behaved as if the other areas did not exist, although they were "suppliers and customers" to each other. The plant manager had set up measures that were area-specific; trying to drive each area to do well. Unfortunately, there were times when doing what was good for the area. For example, increasing production was bad for the downstream process—they couldn't accept the amount produced. Or, the opposite happened, the downstream process was being starved because of planned or unplanned downtime from the supplier, that is, the upstream process. And it was apparent that each of the areas did not communicate their plans or production capability to each other very well. All this also induced variability in the operation of the equipment, frequent startups, shutdowns, diversions of product to intermediate holding tanks, and so on. This induced a high degree of variability and more frequent failures in the machinery and processes.

The plant manager immediately saw the criticality of having a known, consistent production rate across the entire plant for improving production by reducing the variability of the entire process. He immediately established a "pacing unit" to which all the other units produced. When problems arose for meeting this, each area had the responsibility of communicating to the other so that disruptions to the plan could be managed. This approach is similar to the Lean Manufacturing principle of having level flow, and setting a takt time (pacing time) for production. And it's consistent with Lean Manufacturing and Six Sigma, both of which require minimizing variability of the processes for better production capability. While this approach was not called by either name, the solution worked very well; in six months process variability (as measured by the standard deviation of daily production output) was cut in half, and output rose by 20%.

Beta's Martin Plant

Beta's Martin plant was in some serious difficulty for meeting the requirements of a new product launch. The initial analysis had indicated that extrusion was the bottleneck and that they had ample capacity for the additional production requirements demanded by the launch. However, after production began, problems at packaging led to it being identified as the real, or operational, bottleneck. A business-level FMEA was done on the packaging line, identifying the various business-level failures in the packaging system. After the review, the decision was made to clean the line thoroughly (a bit like "detailing" a car, and similar to 5S, or having a "Kaizen event"). While doing the cleaning, the team members were reminded of the various defects and failures identified from the business-level FMEA, and to look for things that might be wrong with the line. Numerous opportunities for immediate corrective action were found:

- The tare weight control was badly contaminated with product, and had not been calibrated in months, resulting in frequent stops and resets, and over-filling.

- The bag magazine had bent loading racks, resulting in jammed bags; photo eyes and reflectors were also misaligned, resulting in inconsistent bag feeding.

- The cutters for the bag tops were different types and settings, and were gummed up with lacquer from the bags, resulting in poor consistency of cuts and rejects.

- The glue pot was set 10° too high, resulting in thin glue and poor sealing, and the glue nozzles were fouled with dried glue resulting in inconsistent application.

- The mechanical "fingers" for spreading the bags were bent, and one had a bad sleeve bearing, resulting in poor filling rates and rejects.

- Control cams were caked over with grease, resulting in misalignment of bags, and frequent jamming.

- Many guides were bent and/or misaligned, and bolts were loose or missing, resulting in inconsistent feed rates in the process.

- Chains were loose, resulting in a jerking motion, and jamming, as the packaging progressed.

- Several bearings were worn out, unlubricated, under-lubricated, or over-lubricated, resulting in jams and inconsistent feed rates.

- The palletizer had photo cells and reflectors misaligned or broken, uncalibrated limit switches, loose chains, and a faulty solenoid, resulting in inconsistencies in the palletizing process and frequent stops.

- The conveyor had poor tracking, poor knockdown bar settings, and poor shrink wrap settings, resulting in inconsistent feed rates.

- Lacquer from the bags accumulated at major wear points resulting in routine jamming of the bags.

Adhering to the business-level FMEA, we also found other issues to be addressed that included:

- Bag specifications and bag quality were a problem. Marketing had insisted on bags of a color and quality that the supplier could not reliably meet, resulting in more frequent jamming of the bags as the packaging progressed.

- Changeovers were a major effort. There were no specific equipment settings for each product and bag type. The changeover/startup consisted of basically running the process until it finally worked, but typically at a rate below design requirements.

- The hopper feeding the packaging machine was undersized for the duty required by this new requirement for production, resulting in inconsistent feed rates.
- There were big differences in practices between shifts for running the packaging line, resulting in inconsistencies throughout the packaging line.

Most of the problems found while cleaning the line were easy enough to address immediately by the team:

- Recalibrating the tare weight.
- Aligning the bag magazine.
- Resetting and calibrating all of the photo cells and reflectors.
- Removing all lacquer buildup.
- Resetting glue pot temperature and cleaning the nozzles.
- Cleaning the cams, and tightening the chains.
- And so on.

Some required a work order to address and assure spare parts were reordered:

- Restoring the bag cutters to like-new condition.
- Replacing all worn bearings.
- Replacing broken solenoids and photo cells.
- And so on.

Other actions still required additional engineering and operational review, for example:

- New operating instructions and PM requirements, including procedures, training, and schedules for all affected operators and technicians. Here, a digital camera was used to take pictures of the packaging line, and then the pictures were used to develop simple instructions and training for operators and technicians to follow. They could actually practice these on the line, using the pictures as a guide. Tactile and visual learning is typically better than reading a manual.

- Resizing the hopper and rates.

- The development of specific settings for changeovers for various products and bag types.

- Working with the bag vendor and marketing to improve performance.

- Setting up an OEE measurement of performance to develop an understanding of losses from ideal.

The result: in just two days, we more than *doubled* production output, and put the plant back on track for meeting the launch requirements, not withstanding the need for still more improvements through better engineering and operating practices.

Beta's Leets Plant

At Beta's Leets plant, serious production problems were being experienced because of routine failures in a device known as "scarfer" (essentially a box-shaped blow torch that "burns" the top layer of a large steel bar to minimize surface cracking). After the initial business-level FMEA to better understand the system-level failure modes, the decision was made to discuss this with the operators of the process. The summary results of this discussion were:

- The scarfer operator complained of not getting straight steel from the "bloom mill," so it "banged and rubbed" the scarfer damaging it and ultimately causing it to fail.

- The bloom mill operator complained of not getting uniform temperature ingots from the "soaking pits" (big ovens) to roll the steel, resulting in inconsistent and uneven rolling.

- The soaking pit operator said the production supervisor told him to leave the tops off the pits to facilitate rapid withdrawal of the ingots, resulting in non-uniform temperature ingots.

Where was the root cause of this problem? Did the supervisor make the right decision to improve production? At this point, we suspect not. But, we still have to do further analysis. What production was being lost because of the scarfer failures that were a direct result of this decision? What increase in production was being achieved because of this decision? Was the trade off appropriate or not? What other systems could be put in place to have faster extractions from the soaking pits and uniform temperature ingots? Were we sure that the additional production from the soaking pits, or not, was consistent with the customer demand, that is, were we applying the takt time principle to level flow and meet demand with minimum inventory and cost? The general conclusion was that the supervisor, while well-intentioned, was wrong in not considering the system-level effects of his decision.

Beta's Van Lear Plant

Beta's Van Lear plant, a steel mill, was working to improve what was already considered good performance. So the decision was made to review the entire plant as a system using the business-level FMEA process. The major elements of the plant were a hot mill, a cold mill, and a finishing mill. Initially the cold mill was considered the bottleneck, and after much debate about including the hot mill in this plant system analysis, the decision was made to exclude the hot mill because it wasn't the bottleneck. We began the analysis, and immediately found that the biggest cause of production loss in the cold mill was the supply—quantity, quality, and consistency of delivery schedule from... the hot mill! It's really important to examine your manufacturing plant as a system, and consider system-level interactions that impact the entire operation.

Beta's Blany Plant

Beta's Blany plant was doing generally well with a notable exception. One of its production lines was under threat of closure and transfer to another plant if productivity did not improve. Costs were high, production was down, and the threat of losing the production line had everyone a bit on edge. An initial review of the production line, a meeting with the production supervisors, and the data on OEE suggested that the front end of the process was the heart of the poor production being achieved.

However, on doing the analysis of the entire production line using the business-level FMEA, the clear conclusion reached was that the packaging section of the production line represented more than one-half the production losses, more than all of the others combined. Individually, each loss was relatively minor, only a few minutes. But, when taken over a week, month, or even year, these small production losses added up to huge losses. You may know that a mere five minutes of production loss per shift in a three-shift, seven-day operation represents a 1% production loss per year. Reduce the number of shifts per week, and the production loss as a percent of the total is even higher.

Most of the problems were similar to those experienced at the Martin plant previously described, that is, just getting the basics right associated with keeping the line clean, and paying attention to the detail of all the critical points in the packaging process. Also critical was getting operators trained so that they can be held accountable for the improvement of the processes, particularly in that area. An intensive effort was put forth by the improvement team, led by the production supervisors in getting these basics right, and in just three short weeks OEE on the production line improved from 50% to 75%. There continues to be much work to be done to inculcate the gains so that they are sustainable, but it appears that the production line will be staying right where it is, for now. The point in the "for now" phrase, of course, is that you can never rest too long, and must always be looking for your next improvement.

Beta's Ivel Plant

At Beta's Ivel Plant, equipment downtime and maintenance, in particular, was being blamed for poor production and high costs. However, the business-level FMEA indicated that the vast majority of the losses were being induced by factors other than equipment downtime:

- The high variability of delivery of the raw material coming into the plant had, by far, the greatest impact on production. The front end process was alternatively starved, or "flooded" with raw material, inducing high variability, the need to move operators from one process to another, and the need to use equipment that could otherwise be mothballed. In Lean Manufacturing terms, flow was not level at all, and indeed was highly variable. Getting more level flow would dramatically improve production, productivity, and costs.

- Oddly, rainy days were the next biggest cause of production loss. It turned out that moisture had a huge impact on the process and high humidity reduced yields and throughput substantially.

- The third biggest cause of production loss was operator absenteeism and/or moving staff from one machine to another to make up a shortfall in a given area. Roughly one-half of this problem would be solved by establishing better flow throughout the plant and eliminating or minimizing the need to move operators from one process to another.

- Equipment downtime was very small compared to these and represented <10% of the total.

Where would you focus your effort, now that you've properly analyzed *system* performance?

Summary

All these case studies have several things in common:

1. It's essential to examine the production process as a system, and fully understand the effects that the different areas within the system have on each other and within themselves.

2. It's essential not to assume that you know where the biggest problem is until you've objectively evaluated the losses in each area and their respective interactions at the system level. Some of the data you have may lead you to the wrong conclusion.

3. It's essential to "go and see" the problems at the shop floor, observing specifically what is happening in the process and equipment. As we will see in the next chapters, Maasaki Imai discusses Gemba Kaizen (or workplace improvement) and

insists that managers "go to Gemba," or go to the workplace to see exactly what the problems are.

It's essential to examine your production system as just that, a complex system with many interactions. A good method for doing this analysis is the business-level FMEA. Its key steps are:

1. Draw a block diagram of the production process.
2. Assemble a cross-functional team from each production area under review.
3. Define a functional failure of the "system" as *anything* that causes loss of quality production capability, or extraordinary costs (you can add other elements like safety or environment events if you wish).
4. Analyze your system-level failures (types, frequencies and effects, potential causes) for each step of the production process and identify those failures that have greatest impact, cost, or quantity of product.
5. Prioritize key processes or equipment for action using the decision-making model shown in Figure 6-3.
6. Next, select those high priority issues that you need to address in light of your available resources.
7. Lastly, select the appropriate tool or strategy to best address these problems.

In these cases, we found that doing a business-level FMEA, resulted in:

- Better teamwork
- Lower failure rates and downtime
- Improved output
- Lower operating costs

As noted, it's essential that we view our manufacturing process as a highly complex and variable business system, and that we understand the failures in that system. Viewing anything that results in a loss of production capacity as a failure in our business system provides a good start for working as a team in quantifying those losses. Quantifying those losses and then applying the appropriate tools on a priority basis will minimize those failures, assuring the success of the business.

However, the tool itself is not sufficient. Management must provide continuous leadership for creating the environment that supports this approach—time, tools, training, measures of success, and sense of teamwork and common purpose. Management must align the organization to that purpose, and must follow up on the findings and action plans, assuring that they are done, measured, and rewarded on a continuous basis. Finally, as shown in Figure 3-2, it's essential that we engage the entire workforce in improvement, not just a few focused on the biggest problems. If we don't resolve the little problems, some of them will become big ones in the future. So, it takes a combination of efforts for world class performance. This model helps to sort out the biggest problems, while we continue to resolve the little ones.

Let's look in more detail at several strategies and tools for manufacturing improvement.

Reference

1.   Moore, R. *Making Common Sense Common Practice: Models for Manufacturing Excellence*, Boston, MA and London, England: Elsevier Butterworth-Heinemann, 2004.

Lean
Manufacturing

7

The Toyota Way can be summarized through the two pillars that support it: continuous improvement and learning (and) respect for people.

Gary Convis

Lean Manufacturing: A Very Brief History

The concept of lean manufacturing first came to be more widely known with the book *The Machine That Changed the World*.[1] It was an excellent review of the history of the development of Lean Manufacturing, primarily at Toyota, but it was scant on the details of the methods for achieving it. More recently, two books—*Running Today's Factory* by Charles Standard and Dale Davis[2] and *The Toyota Way* by Jeffrey Liker[3]—have provided a more clear description of Lean Manufacturing principles and how to apply them. Key points of emphasis from the Standard and Davis book appear to be reducing process variability, reducing system cycle times, and, above all, eliminating waste in the manufacturing process and supply chain, from receipt of order to delivery of product and payment. Liker's book uses a pyramid model to outline Toyota's approach. The foundation of the pyramid is management basing its decisions on a long-term philosophy, even at the expense of short-term financial goals. The next layer in the pyramid relates to having the right processes such that production flow is level, is "pulled," is visual to easily highlight problems, and is standardized. The next step relates to treating people and partners with respect, while challenging and growing them. The final step, or apex, of the pyramid in Liker's model is problem solving using various improvement tools such as Kaizen and Genchi Genbutsu. Liker's model is well founded. If the foundation isn't built on solid ground, all the tools in the world won't help you. Finally, one of my favorites is the "original" book on Lean Manufacturing—*Today and Tomorrow* by Henry Ford,[4] written in 1926, that includes most if not all the same principles outlined in *The Toyota Way*, which was published nearly 80 years later. It would seem that 80 years would be long enough for most companies to implement and sustain these practices.

Several interesting points are made by these authors regarding the history of Lean Manufacturing. For example, following World War II, Japan was very limited in its ability to access resources of all kinds— capital, land, skilled workforce, and so on. These inherent limitations led many of them, and Toyota in particular, to develop techniques to manage a business in a very "lean" environment. These techniques evolved into what is now described as Lean Manufacturing. That is, there was no initial grand strategy for Lean Manufacturing. Rather it evolved as a matter of necessity under the circumstances, initially for survival, but ultimately to their prosperity. Perhaps today U.S. and other manufacturers should take this message to heart. It could be the ultimate survival of the business that is at stake if we don't apply these principles. Unfortunately, most manufacturers have become accustomed to a particular way of doing business, making it very difficult to change the culture of those businesses to apply Lean Manufacturing principles as a matter of habit. One of the more interesting tidbits regarding Lean Manufacturing is that some key practices were developed after Toyota's leaders observed what well-managed U.S. supermarkets were able to accomplish—high inventory turns on small inventories (freshness is critical in food markets), and a high return on capital employed, despite relatively low profit margins on sales. As we'll see in later text, inventory is not considered an asset, although we insist on entering it on our books as such. Rather, it is waste; it is working capital not working, and is subject to damage and deterioration. It also hides all kinds of other problems. Further, in the Toyota model, people are truly assets, ones that appreciate over time through learning and improvement into an exceptional competitive advantage. They are not thought of primarily as costs, something U.S. manufacturers tend to do, where machinery, equipment, and inventory are assets; people are costs. For example, according to Liker[3]:

> "Toyota would never let go or demote workers displaced by productivity enhancements. This shortsighted cost-saving move would create ill will toward the company and prevent all other workers from cooperating in future Kaizen efforts. Toyota always seeks alternative value-added work for workers displaced by production improvements...the worker is the most valuable resource."

A key point that all authors make regarding Lean Manufacturing is that it is not about head count reduction. It is about productivity improvement in the broadest sense of the concept. Costs are viewed as a consequence of practices, systems, and processes. Cost reduction occurs when you improve these. To support this, one of the questions that management should always be asking is, "what can we do to help them (shop floor) do their job better without us?" However, as Imai noted, "Unfortunately, many managers try to reduce costs only by cutting corners; typical actions include firing employees, restructuring, and beating up suppliers."[5] Cost cutting, as we've seen, is a very simplistic approach to cost reduction that is typically not sustainable and should be avoided. Costs are a consequence of your business system design, and it must be continuously improved.

Lean Characteristics

What does it feel like or look like to be lean? What are its characteristics? When you're lean, you *simultaneously* achieve:

- Minimum inventory in the form of raw material, Work in Process (WIP), finished goods
- Minimum product non-conformances, rework, rejects, and returns
- Minimum production losses through unplanned and planned downtime, changeover and transition time, rate reductions and short stops, and quality problems

- Minimum *system* cycle times and minimum delay times between processes
- Minimum variability in production rates and processes
- Minimum unit cost of production
- Excellent on-time delivery performance, customer satisfaction, and gross profits
- Continuing focus and improvement in market share; you're better than you were yesterday, but not as good as you'll be tomorrow!

Optimizing the results from the interaction among all of these variables is exceptionally difficult. It's simply hard work to constantly try to balance these for an optimal business solution. It's also hard work to get the basics in place to assure process and equipment reliability and stability.

Further, in a lean environment you're *less* likely to focus on cost cutting and profit as being:

$$\text{Profit} = \text{Price} - \text{Cost}$$

And more likely to focus on cost reduction (costs are reduced through improved practices and processes), and profit is seen as:

$$\text{Profit} = (\text{Price} - \text{Cost}) \times \text{Volume}$$

You're more willing to trade off lower margins for higher volume and market share, and achieve a higher return on assets over the longer term.

Let's look at more of the details on how to become lean. Consider Figure 7-1,[6] which will be used to illustrate some basic points about Lean Manufacturing. Here, production flow is going to the right, while demand flow is coming from the right. Balancing precisely the variability of production flow with the volatility of demand flow is difficult at best. A lack of product to deliver in

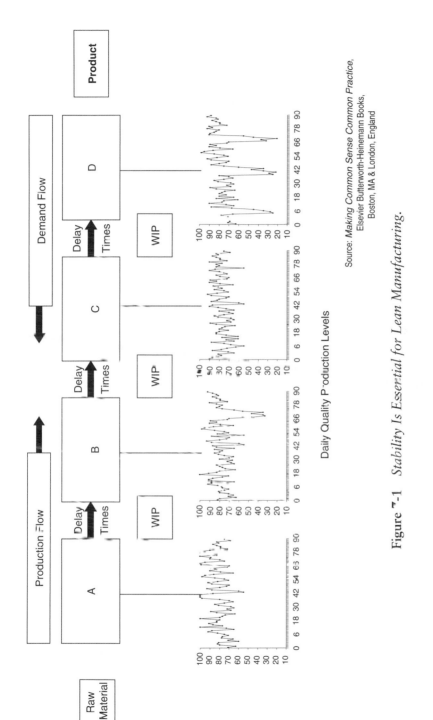

Figure 7-1 *Stability Is Essential for Lean Manufacturing.*

Source: *Making Common Sense Common Practice,*
Elsevier Butterworth-Heinemann Books,
Boston, MA & London, England

a timely way can result in lost sales and profits, and/or unhappy customers who will seek other suppliers. Too much product represents waste in the form of excess inventory; working capital not working. That same inventory incurs additional inherent costs for storage, potential damage, hidden defects, deterioration with time, and so on.

In between processes we often have delay times for one reason or another; for example, it may be that upstream supply has been disrupted because a process has unexpectedly failed, an operator didn't show up for work, or the process isn't working well because of raw material problems. For many of the same reasons, it's also likely that we have inherent variability in each individual process in terms of daily quality output from that process. Top all that off with several different products, or so-called Stock Keeping Units (SKUs), and it results in great complexity in managing while still trying to meet a typically volatile customer demand. We manage all of these delays and variability by keeping buffer stocks—extra raw material, WIP between processes, and finished goods. The greater the variability (un-reliability) in these delays and production output, the greater the need for this inventory. But, this inventory can actually be counter-productive; it hides the problems, as illustrated in Figure 7-2.

A word of caution is appropriate here. Inventory does hide problems, shown here as rocks hidden under a sea of inventory. As will be illustrated in a case study later, several of Beta's plants decided to expose these problems by reducing inventory, to expose the rocks so to speak, only to then run squarely over the rocks, and nearly sink the boat. It's important that we take a balanced view on the reduction of inventory. Some problems are known to us without having to reduce inventory. For example, if production planning changes the production schedule daily, or even by shifts or hourly, to suit sales demands or equipment and process failures, we should not reduce inventory before addressing those problems. We also need level flow in production, a key lean principle. As we address those problems, we will get better stability, and can confidently reduce inventory and still meet demand. Hopp and Spearman offer a method for calculating the amount of inventory and WIP required, given the daily variability and delay times inherent in the system, as well as the confidence limits desired.[7]

Raw Material   Process 1   Process 2   Finished Product
to Customer

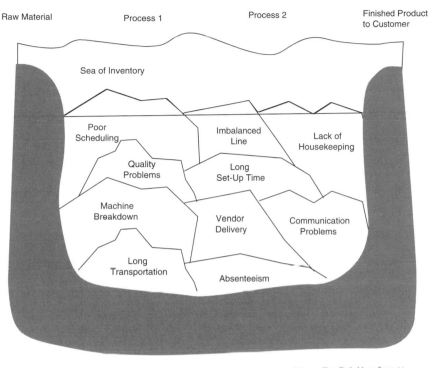

Source: Ron Rath, Marx Consulting,
Grand Rapids, MI

Figure 7-2   *Inventory Hides Problems.*

The Toyota Way[3]

An excellent book on Lean Manufacturing is *The Toyota Way*. Jeff
Liker begins the book with the introduction outlining the information displayed in Figure 7-3, discussed below.

Philosophy: Long-Term Thinking

The 4-P model of *The Toyota Way* covers philosophy, processes,
people and partners, and problem solving. The philosophical
foundation of *The Toyota Way* is long-term thinking, even at the
risk or expense of short-term financial goals and profits. This
seems to be a fundamental reason that most companies fail to
effectively apply Lean Manufacturing principles. Liker indicates

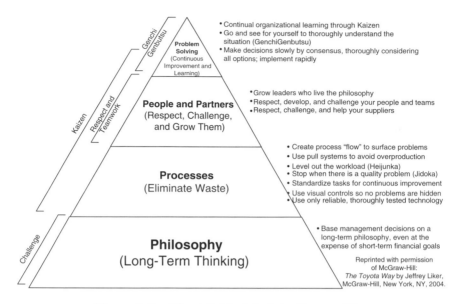

Figure 7-3   *The 4-P Model of the Toyota Way.*

that <1% of companies are truly lean. If lean is such a good thing, and has worked so well for Toyota, why isn't everyone following it? Executives are typically so focused on Wall Street's drive for quarterly results, they are apparently unable (not necessarily unwilling) to behave as long-term thinkers. Lou Gerstner, reflecting on his move from a publicly traded company into a private company, stated[8]:

> "In a private setting, you eliminate the *dysfunctional short-term focus on quarterly results* that dominates the market today... I think there are a lot of executives who are frustrated by the extraordinarily short-term nature of measuring the performance of public companies...The benefit of being a private company is that you have a longer time frame, and you have a direct alignment of the shareholders with the management to fix the company, to build value over time, and be patient with the changes. There's a big difference between being tolerant of poor performance and being intolerant of 90-day numbers without an awareness that maybe [the executives] were investing over that period and doing the right thing."

Jon Huntsman, self-made billionaire and CEO of Huntsman Chemicals, voiced a similar message[9]:

> "Directors regularly make foolish, Wall Street-driven decisions, harmful to the long-term health of the company, because of today's addiction to short-term gains."

Many executives do not appear to have the ability to be strategic in managing the consequences and time scale of the decisions they make today relative to the future of the business in 5 to 10 years, or more. Again, this is not necessarily an unwillingness, but rather an inability because of competing pressures and interests from Wall Street. Toyota apparently has both the ability and willingness to be more long term in its thinking.

Everyone understands the need to make a profit and cash flow the business. It must survive today in order to prosper tomorrow. But, too much emphasis on today's profits leaves the company's success at greater risk tomorrow.

Bob Neurath is concerned that Beta does not have the balance right at this point, and is personally having difficulty adjusting to this relatively "new" way of thinking. Only in the past year or two has Beta has pulled itself back from the brink of its demise. Long-term thinking in that environment was in fact three months, sometimes even day-to-day, as critical decisions were made to assure the company's survival. Now the thinking must change to be longer term. But, Beta is not yet stable enough and, thus, able to think out 5 to 10 years. Perhaps three to five years would be a better time frame. He also notes that he must personally balance long-term thinking with the need for quarterly profits and cash flow, or he won't be around so that he can be long term in his thinking. Moving forward and getting the balance right will not be easy.

Processes: Eliminate Waste

The next layer in Liker's 4-P pyramid calls for managers to understand all of their processes, not just the ones related directly to production. Here, the focus is to eliminate all forms of waste. Waste is defined as anything that adds cost, but does not add value. Examples of waste include the following areas:

- Excess inventory
- Overproduction: in excess of customer or downstream demand
- Motion, movement, and transportation of inventory
- Processing or cycle times
- Unnecessary waiting, motion, movement, and checking
- Defects, rework, rejects, and scrap

And, as Liker appropriately points out:

- Unused employee creativity is a huge waste

According to Liker, the concepts supporting the elimination of waste are numerous:

- Create a smooth "flow" of production to surface problems: reduce the variability of your processes
- Level out the workload
- Use pull systems to avoid overproduction: make products based on customer demand at your takt time—the pace at which the plant operates to just meet market demand (and no more)
- Stop when there is a quality problem
- Standardize tasks for continuous improvement (standard work and innovation are two sides of the same coin—standardize, then improve/innovate; standardize, then improve/innovate in one continuing effort)
- Use visual controls so no problems are hidden
- Use only reliable, thoroughly tested technology

A key tool for understanding your processes is to do value stream mapping for each of your product lines. In this process you symbolically "bolt yourself" to a chunk of raw material and follow it from raw material to finished goods delivered to the customer. Along the way you look for non-value add activities (activities that add cost, but do not add value). You then work to remove non-value adding activities, and add more value adding activities. Although simple in concept, it is much more difficult in practice.

Bob Neurath has had his staff do some preliminary estimates of value add time at several plants. The numbers were astonishing. At one plant, one of its common products had one hour of actual value-add time. But, delivery time from order entry to product delivered was 30 days. Value-add time was one hour of 720 hours. Actual production time, from raw material into the first step through to product being placed into the warehouse was more like 1:20—one hour of value-add in a 20-hour production cycle. This ratio was also typical for many of Beta's products. Clearly, the opportunities were enormous for improving system cycle time and reducing waste.

People and Partners: Respect, Challenge, and Grow Them

The focus here is to:

- Grow leaders who live the philosophy
- Respect, develop, and challenge your people and teams
- Respect, challenge, and help your suppliers

Bob concurs with another of Jon Huntsman's admonitions[9]:

> "The surest path to success is one where others walk with you. Plants and equipment can be replaced easily; hard-working, loyal employees are as valuable as precious gems. They are critical to any leadership success. If CEOs are the soul of the organization, employees are the heart."

Bob's concern here is that Beta still has several leaders who are bound to the old management model of simply surviving through cost cutting. Obtaining better alignment of the organization to a lon-ger-term way of thinking will be crucial for the success of Beta. He reflects on Jim Collins admonition to "get the right people on the bus, in the right seats." It's likely that some of his management team will need to change seats, or buses.

Respecting, developing, challenging, and helping your people and your suppliers is also a double-edged sword. The CEO and his man-agement team must be demanding enough to challenge people, but not so demanding as to be unrealistic such that people don't even try. There are managers who say that you should set big, audacious goals. To Bob this is like setting yourself up for failure. It might work in a crisis for a short period of time before burnout; it might work with a new technology or process that must be ramped up rapidly. But, it's not likely to work long term, and Bob's new view is that Beta is in a marathon, not a sprint. They've been through the sprint, have survived, and now need to re-focus on the marathon, maybe going for a middle distance approach first. He also believes that people must be given the time, tools, and training, along with the challenges to help them develop and apply their skills. This is something that continues to need improvement at Beta. Likewise, squeezing your suppliers for every penny may not be the best approach either. Having supplier-partners seems a better approach and Beta wants to move toward having a genuine partnership with its key suppliers.

Problem Solving: Continuous Improvement and Learning

The apex of Liker's model is problem solving. He quickly admonishes those who see the Toyota production system and immediately begin to try to apply Kaizen events, or do their ver-sion of TPM, or cut back on inventory. That is, companies who try to use the tools before getting the foundational elements in place will be sorely disappointed in their experience with Lean Manufacturing. It begins with a philosophy of long-term think-ing, moves on to understand your processes and the waste in them, then on to respect and challenge your people and supplier partners. The last step in *The Toyota Way* is using the tools. Yet so many companies go with the tools first. As the old adage goes, it's like putting the cart before the horse, and that system doesn't work very well.

Bob's opinion is that Beta seems to be putting the cart before the horse too often. It seems there is too much focus on tools, and not enough on the foundational elements. As we observed in Chapter 1, Beta has had mixed success in applying the various improvement tools and strategies; some plants have been successful, others have not. Even the successful ones seem to have difficulty sustaining the gains, or getting beyond a level at which they stall. His conclusion is that more long-term thinking needs to be done, and processes must be better understood to better eliminate waste and do more value-add. On the point of respect for people, he's ambivalent. Certainly he respects Beta's people and wants to challenge them to be more productive, if he provides additional training and tools. However, the company's overall track record here is spotty. Beta had to lay off several hundred people during its long climb out of the proverbial hole. This is clearly contrary to *The Toyota Way*. Unfortunately, given their business situation, it was a necessity. Beta had to cut costs in some business units in order to survive. Now the question is how do we reinvent those business units and show a deep respect for people, while still challenging them to improve.

Another concern regarding the Toyota model is the make to demand or Kanban thinking. While he can conceptually grasp this principle as working reasonably well at Beta's discrete plants, it looks more difficult at its process plants. The process plants tend to push production versus responding to pull; they do not cycle well, that is, ramp up and down to match demand. For example, getting a refinery or a smelter working properly takes considerable time, and the option of running it much below its design rate, without incurring substantial yield and other penalties does not exist; shutting down brings on other risks and difficulties. The good news is that they generally have a level production flow. Perhaps if the time frame for this thinking were extended, this approach would work better. That is, for the discrete plants the time frame for Kanban might be hours, days, or weeks, depending on the process under consideration, but for process plants the time frame might be weeks or months. In either event, forecasts and customer order management become even more critical to assure level flow while minimizing inventory.

Keeping Things Simple

Liker indicates that Toyota shuns complex tools like Six Sigma, preferring to keep things simple. The key focus in problem solving is:

- Continual organizational learning through Kaizen
- Team members using four key tools:
  - Go and see for yourself at the workplace to thoroughly understand the situation or problem
  - Analyze the situation and establish counter-measures for immediate problem resolution
  - Then ask "why?" five times to get to the root cause; once a root cause is established, standardize the solution (illustrated in Chapter 13)
  - Use one-piece flow and Andon (signals/lights) to surface problems
- Making decisions slowly by consensus, thoroughly considering all options; and then *implementing rapidly*

Pareto diagrams are used to highlight major problems and analyze them. The proverbial 80/20 rule applies: 80% of your problems or losses are driven by 20% of your equipment or processes. Pareto diagrams allow for quick focus on the major problems.

Improvement plans are also kept simple, and limited to an A3 (~11" × 17" paper size) presentation. These plans typically describe:

- Current situation
- Proposal; labor cost and time analysis
- Future situation/result
- Plan, implementation, controls, time line (i.e., Plan, Do, Check, Act)

An example of an A3 strategy document for one of Beta's operations is shown in Figure 7-4.

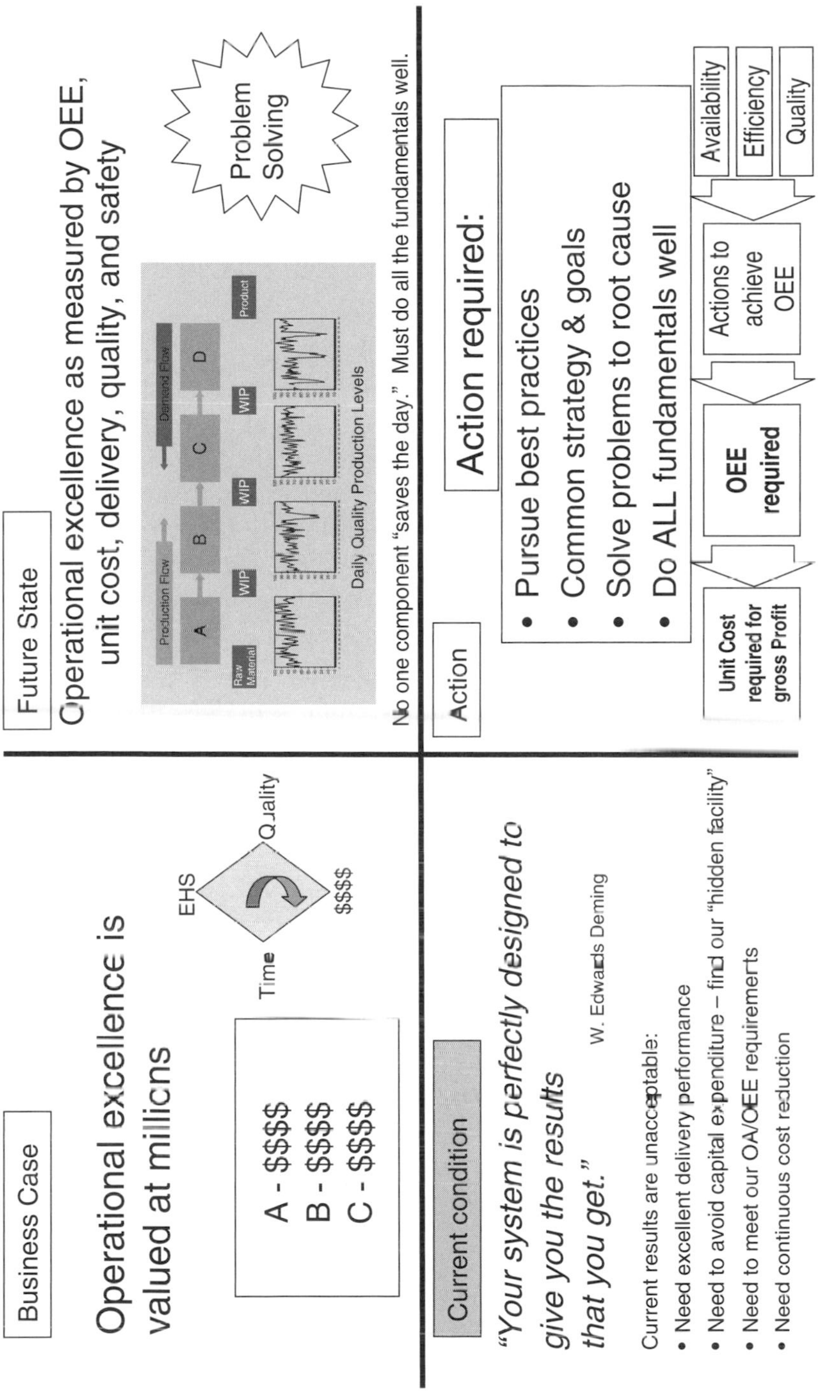

Business Case

Operational excellence is valued at millions

A - $$$$
B - $$$$
C - $$$$

EHS

Time

Quality

$$$$

Future State

Operational excellence as measured by OEE, unit cost, delivery, quality, and safety

Problem Solving

Production Flow → Demand Flow

Raw Material | A | WIP | B | WIP | C | WIP | D | Product

Daily Quality Production Levels

No one component "saves the day." Must do all the fundamentals well.

Current condition

"Your system is perfectly designed to give you the results that you get."

W. Edwards Deming

Current results are unacceptable:
- Need excellent delivery performance
- Need to avoid capital expenditure – find our "hidden facility"
- Need to meet our OA/OEE requirements
- Need continuous cost reduction

Action

Action required:

- Pursue best practices
- Common strategy & goals
- Solve problems to root cause
- Do ALL fundamentals well

Unit Cost required for gross Profit

OEE required

Actions to achieve OEE

Availability
Efficiency
Quality

Figure 7-4 *Example A3: Strategy for Achieving Operational Excellence.*

As we'll see in later chapters, process and equipment stability and reliability are facilitated by the use of:

- Kaizen, including 5S and TPM (Overall Equipment Effectiveness [OEE], operator care, Preventive and Predictive Maintenance, training)
- Standardized work and process consistency/conformance

Finally, and perhaps most importantly, *management stability is a must*. Toyota's promotion and management changes are slow by U.S. standards. It's common for a manager to be in a given role for 5 to 10 years. Their view seems to be that it's hard to have excellent process stability and organizational alignment when the management team changes frequently. In Chapter 5 we discussed how management instability (changing every two years or so) was a major factor in the dramatic decline in performance of Beta's Omega division.

And, according to Liker, Lean Manufacturing can also be counter-intuitive. Given the aforementioned guidelines, it may surprise some to find that in the Toyota Production System, it may be best:

- To selectively build up inventory, if that improves leveling of flow and workload
- To add overhead support, if that adds more value and reduces waste in the process
- Not to keep workers busy, if that results in over-production or excess inventory
- To use manual processes over automation, if you haven't optimized the manual process; and manual can be more flexible for future needs
- To idle a machine or stop producing: overproduction is waste, especially if quality is poor

Toyota Production System Summary

The Toyota production system is differentiated from *The Toyota Way,* and depicted in Figure 7-5.

Reprinted with permission of McGraw-Hill from *The Toyota Way* by Jeffrey Liker, McGraw-Hill, New York, NY, 2004.

Figure 7-5 *The Toyota Production System.*

According to Liker, the Toyota Production System is NOT:

- A tangible recipe for success
- A management project or program
- A set of tools for implementation
- A system for the production floor only
- Implementable in a short- or mid-term period

The Toyota Production System is:

- A consistent way of thinking
- A total management philosophy
- Focused on total customer satisfaction
- Focused on teamwork and improvement
- A neverending search for a better way

- A system where quality is built into the processes
- An organized, disciplined workplace
- Evolutionary

Decoding the DNA of the Toyota Production System[10]

A slightly different, but compatible view of the Toyota production system is taken by Spear and Bowen.[10] Similar to Liker, they advise that it is not so much about the tools, as it is about their processes and culture. One paradox they observe is that the rigid specifications that Toyota demands for all its processes is the very thing that makes it flexible and creative. Whenever Toyota defines a specification, it is establishing sets of hypotheses that can then be tested, following the scientific method. Surprisingly, this rigid approach does not result in a command and control management style that one might expect, because it is pushed to the lowest level in the organization, and throughout. They report that the system grew naturally over five decades and has never been written down, making it difficult to actually articulate or define. Their view is that it works because of four rules that provide the tacit knowledge inherent in the system:

1. All work shall be highly specified as to content, sequence, timing, and outcome.
2. Every customer-supplier connection must be direct, and there must be an unambiguous yes or no way to send requests and receive responses.
3. The pathway for every product and service must be simple and direct.
4. Any improvement must be made in accordance with the scientific method, under the guidance of a teacher, at the lowest possible level in the organization.

Toyota workers learn the rules by doing the work, and problem solving. Workers are asked:

1. How do you do this work?
2. How do you know you are doing this work correctly?
3. How do you know that the outcome is free of defects?
4. What do you do if you have a problem?

These simple questions provide the worker with the understanding and ultimately the skills to do a better job.

Surprisingly, they report that Toyota does not consider the tools or practices, such as Kanban or Andon, as fundamental to the Toyota Production System. Rather, these are temporary responses or countermeasures that serve to maintain production while permanent solutions are found. They use all of these approaches as they strive for the ideal production environment.

Beta's Banner Division

On reviewing Toyota's model as it was outlined by Jeff Liker, Bob Neurath had a number of concerns regarding Beta's use of Lean Manufacturing and the Toyota Production System in its Banner division. As previously noted, Liker admonishes companies that Toyota Production System is not a set of tools that can be quickly implemented, nor is it a short- or mid-term solution to any company's problems. In Liker's pyramid model, he also admonishes that companies must first take a long-term view even at the risk of short-term profits, then understand its processes, then respect and challenge its people and suppliers, and the last step is to apply the problem-solving tools. Granted there may be some overlap and/or feedback loops between layers in the application of the model, but this is the general priority.

A review of Banner's performance against this model suggested that they seem to have things in reverse, that is, they have a strong focus on applying the tools without getting more of the foundational elements in place, and part of that was his doing. Beta could little afford to have a long-term view, when the company was in a difficult financial position. You have to survive today to prosper tomorrow. Now it was time to take a longer-term view. Banner had been applying its version of Toyota Production System for several years now, with little apparent impact. While they hadn't gotten worse in financial performance, they hadn't gotten much better either. Bob was also concerned that a focus on using the tools to solve immediate or obvious problems had led to a situation where insufficient focus had been given to having a comprehensive understanding of their processes and where the bigger opportunities might be. A better understanding of their processes through value stream mapping or using the business-level Failure Modes and Effects Analysis (FMEA) would likely be a better approach. Further, Banner had a number of layoffs as part of its restructuring effort to survive, leaving most employees demoralized. It would be difficult to gain their respect and trust in an environment where they had suffered through layoffs. Challenging them to "work harder" did not seem to be the answer either. To the people on the shop floor, the reduction was due to poor management, not poor workers. Right or wrong, that was their perception, and it had to be overcome. These more fundamental issues needed to be addressed. Several examples are provided in the following text on Banner's use of Lean Manufacturing principles.

Banner Division's Boldman Plant

At the Boldman plant, the management team had been admonished that inventory was waste, and so they diligently began to reduce inventory in one of the production lines that fed a critical component into the overall production process. A business-level FMEA of this production feed line concluded that a lack of inventory was inducing another $250,000 in additional costs because of overtime for unplanned production, loss of production in the final production process, and other miscellaneous impacts. The opportunity cost saved from reducing the inventory was some $25,000. Apparently the Boldman management did not understand some of the finer, more counterintuitive points of Lean Manufacturing that said it may be necessary to build up inventory in the short term, if the additional inventory helps level flow and improve overall workload and production, while you solve the longer-term problems.

Banner Division's Blue River Plant

At the Blue River plant, a Kaizen event was performed to improve a particular machine that was viewed as critical to the production process. The event was completed, and the machine's performance improved substantially. However, while the event was being performed, the following observations were made:

Flow. Even before considering use of the tools to solve problems, someone should have examined the process flow to determine if this was the right machine for the Kaizen event. Considering its condition, the machine actually worked reasonably well, and appeared to be operating at about 90% of its capability, which would have supported production flow requirements. Most of the production loss was occurring because of a lack of steady flow through the machine from the upstream process, although a large backlog of raw material was waiting there to be processed and sent to this machine. This apparently related to the upstream process being optimized, rather than optimizing the entire production process. Other problems related to flow had to do with the shift handover process, that is, the machine was routinely *not* set up for the next shift to run, resulting in a one to two hour delay in production between shifts. A much better understanding of process flow was needed, and meeting "customer demand" at the next process downstream, another lean principle, was essentially being ignored. And, at times, the upstream process appeared to be overproducing, and not leveling flow.

5S, Kaizen, TPM. The machine was relatively new, that is, only three years old, but it was in very poor condition. When it was new, it had met all its operational requirements. Unfortunately, the 5S principles and proper housekeeping that are fundamental to Lean Manufacturing had not been done, nor had basic preventive maintenance, that is, also fundamental to Lean Manufacturing (i.e., tighten, lubricate, clean, and calibrate the equipment on a regular basis). Associated with this, leaks were abundant, and several components were actually missing. Alarms were largely ignored (reset or bypassed). Several components were coated with dirt and were unrecognizable as to their function. Another fundamental of Lean Manufacturing was being ignored—make problems easy to see and correct. Essentially all maintenance was done on a fix-on-failure basis. Most of these problems could have been addressed by having a good PM program to regularly clean, calibrate, lubricate, seal, and so on, which included having operators do certain cleaning, adjustments, inspections, and lubrications. The machine needed to be restored to a like-new condition and performance or better, a TPM principle, and then kept that way. While this was a Kaizen event, fundamental elements of Kaizen, 5S, and TPM, which we'll learn more about in later chapters, were being ignored up until the Kaizen event was conducted. This was NOT a proper application of Kaizen. This Kaizen event was mostly needed because Kaizen principles had been ignored to that point.

Standard Work. Raw material was coming from the upstream process from two different production areas, each of which was providing material made to obviously different specifications, although each used a very similar process to make the raw material for this one. Both production areas upstream of this process needed to be advised to apply common standards for the raw material supplied. Of course, standard work is another fundamental principle for Lean Manufacturing, and it was not being followed.

Safety and Health Issues. Another concern related to the overall poor housekeeping was the need to minimize the risk of safety or health incidents. Debris was abundant—dirt, hoses, cables (e.g., trip hazards; walkways were obscured by dirt; a safety rail was missing). Mixed signals were abundant: we're saying we care about safety, but let our people work in this kind of environment.

Respect for People. Beyond the implications of health and safety issues regarding respect for people, the plant had recently had a layoff and those remaining were being threatened with additional layoffs if the problems continued: "the beatings will continue until morale improves."

There were more examples like this that were similar for other Banner division plants. Bob had sufficient information to know that Beta's Banner division did not understand Lean Manufacturing. They had gone immediately to the apex of the pyramid, without addressing the fundamental issues first. He was determined to make sure this was corrected.

Summary

Lean Manufacturing begins with the foundational element of having a philosophy of long-term thinking, even at the risk of short-term financial goals. Short-term financial goals cannot be ignored, but Beta must give greater deference to long-term thinking and goals. This will continue to be difficult for Beta given Wall Street's demand for quarterly results.

Getting the basics right and having reliability and stability in the processes and equipment is a must. If you don't have reliable processes and equipment, it will be very difficult to be lean. You need that extra "stuff"—buffer stocks, spare parts, spare equipment, and so on—to manage your **un**-reliability and still meet your customer's demands.

Beyond this, the processes and flow must be better understood and steady flow must be established to match customer demand. This will be more difficult for Beta's process plants, which are much better designed to "push" production, although with more level flow. Longer-term thinking may be necessary here to match the process to the demand, and certainly better forecasting of demand is needed. Given their recent history of layoffs, simultaneously respecting and challenging people will be difficult for Beta as it seeks to re-establish trust with the workforce. Thinking long-term should help here, and the forthcoming loss of baby boomers to retirement may help Beta bring in new blood and establish a more open, trusting culture. At the same time, much better training, and a learning environment must be created to support the new people. The problem-solving tools have been pushed too hard, to the detriment of the other foundational elements. Bob will give greater emphasis to the foundational elements, and expects to use the tools better, giving the people the time, tools, and training, so that they perform a better job. Doing this consistently in the coming months and years, and having "constancy of purpose," should help shift the culture of the organization to better apply Lean Manufacturing principles.

References

1. Womack, J., Jones, D.T., and Roos, D., *The Machine That Changed the World, The Story of Lean Production*, New York: Harper-Collins, 1991.

2. Standard, C., Davis, D., *Running Today's Factory, A Proven Strategy for Lean Manufacturing*, Cincinnati, OH: Hanser Gardner Publications, 1999.

3. Liker, J., *The Toyota Way*, New York: McGraw-Hill, 2004.

4. Ford, H., *Today and Tomorrow*, New York: Productivity Press, 2003; original by Doubleday, Page & Company, 1926.

5. Imai, M., *Gemba Kaizen*, New York: McGraw-Hill, 1997.

6. Moore, R. *Making Common Sense Common Practice: Models for Manufacturing Excellence*, Boston, MA and London, England: Elsevier Butterworth-Heinemann, 2004.

7. Hopp, W., and Spearman, M., *Factory Physics*, Chicago: Irwin Times Mirror, 1996.

8. Thornton, E., *Going Private*, Interview of Lou Gerstner, former CEO of IBM, Business Week, February 27, 2006.

9. Huntsman, J., *Winners Never Cheat*, Upper Saddle River, NJ: Pearson Education, Inc., 2005.

10. Spear, S. and Bowen, H.K., *Decoding the DNA of the Toyota Production System*, Boston: Harvard Business Review, September-October 1999, reprinted with permission from Harvard Business School Press, copyright 1999 by Harvard Business School Publishing Corporation, all rights reserved.

Kaizen

8

Today's best, which superseded yesterday's, will be superseded by
tomorrow's...

Henry Ford

Introduction

Kaizen® is a Japanese word meaning change for the better or continuous improvement. It is fundamental to Lean Manufacturing that
we must continuously strive to get better. In this chapter we'll cover
the basic concepts of Kaizen taken from Gemba Kaizen® by Masaaki
Imai,[1] and we'll review how Beta is applying those concepts.

Interestingly, Henry Ford understood and described many of the
concepts of what is today known as Lean Manufacturing and Kaizen
some 80 years ago when he wrote *Today and Tomorrow*. Indeed,
Toyota at one point gave its American managers a copy of *Today
and Tomorrow* to read.[2] U.S. manufacturers, including Ford Motor
Company, could take a lesson from Henry Ford's "playbook,"[2]
which parallels many of the concepts in *The Toyota Way* by Jeffrey
Liker.[3] In it, among many other points Ford makes, he observed that
"Today's standardization...is the necessary foundation on which
tomorrow's improvements will be based." Like Toyota, he thought
of standardization and innovation as two sides of the same coin.
And, like Toyota he abhorred all forms of waste,. He applied Lean
Manufacturing and Kaizen principles at every opportunity, seeking

constantly to reduce waste, reduce variability, reduce system cycle times, and improve overall performance.

Imai provides an outline for continuous improvement and adds much detail to Henry Ford's musings, providing much of the "how to" in implementing Kaizen in the workplace. Early in the book he observes that:

> "Western management . . . worships innovation; major changes in the wake of technological breakthroughs; the latest management concepts or production techniques. Innovation is dramatic, a real attention getter. Kaizen on the other hand, is often undramatic and subtle. But innovation is one-shot, and its results are often problematic, while the Kaizen process, based on common sense and low-cost approaches, assures incremental progress that pays off in the long run. Kaizen is also a low-risk approach. Managers can always go back to the old way without incurring large costs."

In Imai's view, focusing on the "next-big-thing" is a high-risk, expensive approach. Kaizen is inexpensive, low-risk, and subtle. Day-to-day improvement tends to be fairly low profile, with little glamour. As we discussed in Chapter 3, Kaizen is what could be called "little" innovation. And, as we discussed in Chapter 3, there is no prohibition to doing both. Indeed, a strong case can easily be made for doing both. It's the little innovations that engage the entire workforce, lower your manufacturing costs, provide higher gross profits, and create subtle competitive advantages that others will have great difficulty in copying. Lower costs also fund the "big" innovation that will help find major advances and more dramatic improvements. Creating an environment where innovation is the norm throughout the organization is the best of both worlds. Before going into Kaizen principles and practices, let's first consider the suggestions of Robinson and Schroeder, which are highly consistent with those of Imai.

Ideas Are Free

In their book *Ideas Are Free*, Alan Robinson and Dean Schroeder[4] make a strong case for Kaizen, although they do not refer to the principles they review as Kaizen. They observe that a typical response to flagging profits and increasing competition is cost cutting and lay-offs, when all the while help is close by; the people doing the work see many, many opportunities for improvement. Taking advantage of their knowledge of the problems, and more importantly the potential solutions, is essential to a company's continuing success. Margaret Wheatley,[5] in a similar vein, observed that "People own what they create." It is really important to have people express their ideas and creativity; they will own the solutions to the problems they solve, and are much more likely to apply those solutions, and continue to look for even more improvements. Such a simple, common sense approach by whatever name it is called should be more widespread.

Robinson and Schroeder go on to make several more observations and recommendations, which are summarized below.

Knowledge is characterized as *aggregate*—usually from management (e.g., sales are down 10%; profits are down 20%, and so on) and *detailed*—knowledge of the circumstances of time, place, and know-how. In large organizations of thousands of employees, managing complexity requires ideas and solutions be driven to the lowest level of the organization to help manage change in people, in practices, in customers, and so on. Managers only see a fraction of the problems (disguised opportunities), and the more complex the system, the less they see. Hence, they must rely on the people doing the work to address the complexity and provide detailed problem solving. It is impossible to achieve excellence without attention to detail, and that must come from the people doing the work and their many small ideas. The bigger the organization, the more important it is to welcome ideas and solutions from the lowest level in the organization.

They are very skeptical of the classical ideas program. It is often daunting and counterproductive. In their view, big ideas or the latest advance is more likely to be countered or copied. Rather, lots of small ideas that are successfully implemented really give one a competitive advantage, one that is very difficult to copy. These small ideas tend to remain proprietary, accumulating over time into sustainable competitive advantage, like the classical flywheel effect! These ideas assure attention to detail; the best companies do the basics really well. And as a bonus, lots of people are engaged in finding and developing the little ideas and creating an environment that increases the probability of finding big ideas! An innovative shop floor creates a flexible, responsive, and adaptive culture where change and improvement are routine.

Regarding rewards, they are *not* supportive of big incentives in a traditional ideas program. Indeed, they argue effectively that establishing a reward program for ideas can actually be a "*dis*-incentive" to idea generation because:

- The benefits and value is often hard to calculate.
- Only ideas that are associated directly with money are submitted.
- Rewards go to those who submit ideas, not the myriad of people who implement them.
- Big rewards can encourage undesirable behaviors, such as:
  Stealing of others ideas and cheating to get the rewards.
  - Managers being incentivized to try to reduce the rewards; the money comes out of their budgets.
  - Ideas are withheld until a reward is possible (e.g., not in the design, but later when a greater reward might be possible).
  - Staff manipulating the system for personal gain.

As shown in Table 8-1, they offer data supporting the principle that rewards don't improve idea generation or results.

The data in Table 8-1 can be summarized as lots of small ideas have a much greater financial impact. And, the more money a company offers for ideas, the fewer ideas it receives. The most effective reward is that your idea was used! Such a system requires exceptional trust in the employees. Most if not all of your people want

Table 8-1 *Effectiveness of Idea Programs*

	Japan	US
New ideas per employee	37.4	0.12
Participation rate	77.6%	9%
Adoption rate	87.3%	32%
Net savings per idea	$126	$6,114
Net savings per 100 staff	$422K	$23K
Average reward per idea	$2.83	$602

to do a good job. Trust them to do one, challenge them to improve by solving problems at their level of responsibility and they will and will be proud of their work, and the business will do better. They say that rewards for improvements should be based on high level, aggregate measures reflecting collective performance. Benefits are distributed to *all* employees, equitably, using clear, public rules. Ideas and rewards are integrated into the way the company is routinely managed.

In their view, an effective idea system has the following attributes, which are consistent with *The Toyota Way* and Kaizen principles:

- Ideas are part of everyone's normal work.
- Ideas are easy to submit.
- Ideas are reviewed by those with direct knowledge of the work and its impact, and can suggest improvements to make it work.
- Decision making is rapid, effective, and efficient.
- Feedback to each suggester is quick and complete.
- Implementation is rapid.
- Ideas are reviewed for additional potential.
- People are recognized and success is celebrated (typically in small ways).
- The system and results are actively measured, managed, and improved.
- Senior management must provide the systems and leadership for its success.
- Innovation and improvement must permeate the organization *at all levels.*

Whether we think of the improvement process as Kaizen, as an ideas program, or by any other name, clearly the people who are closest to the work have more knowledge of that work and must be involved in making the work more efficient and effective. Creating an environment where innovation is routine throughout the entire organization is essential for continuing business success. Let's return to principles espoused Imai in *Gemba Kaizen*.

Major Principles of Kaizen

> "As a general rule of thumb, introducing good housekeeping in Gemba reduces the failure rate by 50%, and standardization further reduces the failure rate by 50% of the new figure."
>
> *Masaaki Imai*

Could this statement possibly be true? This represents a 75% reduction in the failure rate by just doing good housekeeping and standard work. If it were true, why wouldn't all manufacturers be tending to their housekeeping and standard work? Imai appears to know of what he speaks, and Toyota has benefited immensely from Kaizen principles. But let's suppose Imai is wrong, and these percentages are really only 25% and 25%. Even if he's only half right, it's still a major improvement of some 37.5%. And my experience has been that housekeeping in most plants is fairly shabby. And most plants performance is likewise, fairly ordinary. Let's assume for the sake of argument that housekeeping in itself may not provide lower costs. What it will do, however, is instill a sense of pride and craftsmanship that permeates the organization. When you live in a pig sty you behave like a pig. Taking pride in your workplace, and taking care of it is a first order of business. According to Imai the major focus of Kaizen must be housekeeping, muda (waste) elimination, and standardization. This harks back to the key concepts of the Toyota production system, but then Kaizen is an essential element of it.

Kaizen and Management

The primary goal of Kaizen is the simultaneous achievement of excellence in quality, cost, and delivery:

- *Quality* of finished products, of intermediate products (work-in-process), and of the processes for making the products. Poor quality must *never* be passed along to the next process. Total Quality Control and Management (TQC and TQM) are key elements supporting quality production.
- *Cost*: the overall cost of designing, producing, selling, and servicing, not just the manufactured cost.
- *Delivery* of the requested volume on time, all the time

All three conditions must be met to satisfy customers and support the business.

Imai **often** stresses the importance of management "going to Gemba" (the shop floor) to understand the processes and problems, then to take temporary countermeasures and beyond to routinely find the root cause of the problems using a simple 5 Whys method. Once the problem is truly solved, management must assure that standards are established or revised, and that people are trained in and work consistently to those standards to prevent future re-occurrence.

Management must remove the obstacles for good Gemba by instilling in people a sense of pride and mission, and aligning the organization to its mission. Imai observes that shop floor needs (problems) are best known by the people on the shop floor and, hence, must be solved by them. Doing so minimizes the resistance to change ("people own what they create"), and people look forward to offering even more solutions.

Major Kaizen systems include:

- Total quality control/total quality management
- A just-in-time production system (Toyota Production System)
- Total Productive Maintenance (TPM)
- Policy deployment
- A suggestion system (see "Ideas Are Free" section)
- Small-group activities (see "Teams and Teamwork" in Chapter 4)

The three major activities of Kaizen are:

- 5S—sort, straighten, scrub, systemize, standardize
- Standardization—the first objective of management is to maintain standards
- Elimination of muda (waste). For example:
  - Overproduction
  - Inventory
  - Repair/rejects

- ▪ Motion
- ▪ Processing
- ▪ Waiting
- ▪ Transport

Each of these is discussed below.

5S

The key features of 5S are:

- ▪ Sort—separate all the unnecessary parts and tools and eliminate them.
- ▪ Straighten—put things in order for easy access
  - ▪ You must need it in next 30 days if it is nearby.
  - ▪ Use shadow boards and other organizing tools.
- ▪ Scrub—clean everything.
  - ▪ Clean to inspect
  - ▪ Inspect to detect
  - ▪ Detect to correct
- ▪ Systematize—make cleaning and checking routine.
- ▪ Standardize—the above steps to make it continuous.

5S is considered a prerequisite for TPM.

Standardization

In Imai's model, the key features of standardization are:

- ▪ Standards reflect the *current best, easiest,* and *safest* way to do a job
- ▪ The best way to preserve know-how and expertise

- Standards show the relationship between cause and effect
- Standards provide:
  - a way to measure performance
  - a basis for maintenance and improvement objectives
  - a basis for training
  - a basis for audits and diagnosis
  - a means of preventing recurring errors and variability

Elimination of Muda (Waste)

In Kaizen, cost management is not cost cutting. *You reduce your costs by reducing waste* through activities that continuously improve productivity and quality, and reduce waste in the forms noted. Imai opines that most manufacturers use four times as much space, twice as many people, and 10 times more lead time than they really need, creating huge amounts of waste and extra costs.

In a Kaizen environment: (1) management must facilitate innovation and improvement, and (2) the shop floor must support improvements and maintain them. It might also come as a surprise that in an organization using Kaizen properly, the organization is much more process-oriented, and much less results-oriented. The well-founded belief is that if you have the right processes, you will get the right results. Focusing only on results is less likely to be sustainable over time. In doing Kaizen, you must apply the following.

- Follow the Plan, Do, Check Act (PDCA) cycle (a Deming principle from decades ago, and one similar to the Six Sigma concept of Define, Measure, Analyze, Implement, Control [DMAIC])
- Follow its close cousin: Standardize, Do, Check, Act Cycle (SDCA)
- Put quality first: Quality, Cost, Delivery are three simultaneous goals—quality is always first
- Speak with data: collect, verify, analyze, act (similar to Six Sigma tools)

- The next process is the customer; never pass on defective parts or inaccurate information

Common phrases in a Kaizen environment:

- Gemba—the workplace and the shop floor—management must spend a lot of time there.
- Muda—waste of all forms—eliminate it!
- Mura—variability or irregularity—minimize it!
- Muri—difficulty or strain—minimize it!
- 5S—Seiri, Seiton, Seiso, Seiketsu, Shitsuke (sort, straighten, scrub, systematize, standardize)—just do it!
- 5M—Manpower, Machines, Materials, Methods, Measurements—align them to a common strategy and purpose
- Poka Yoke—mistake proofing—make it easy to do the job right
- SMED—single minute exchange of die, or quick changeover.
- Takt time—the pacing of the production process to the theoretical time it takes to make a single unit of product for a customer (i.e., divide total production time by the number of units; use takt time to pace production to just meet demand).
- Kan Ban—a "sign board" that prompts "re-order" supporting a pull system and responding to customer demand, or the next step in the production process—make what is needed, when it is needed (just-in-time), and again, never pass along poor quality product to the next step.

Kanban

The Kanban system is illustrated in Figure 8-1.

As shown in Figure 8-1, customer demand "pulls" supply into the market and at each preceding step in the production process. Kanban cards are used to signal the upstream process when to supply more. Production is paced according to the takt time for just meeting

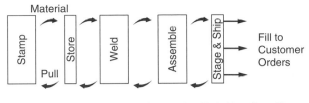

Source: Ron Rath, Marx Consulting,
Grand Rapids, MI

Figure 8-1 *Kanban Illustration: Pull Production System.*

customer/downstream demand. A key point, however, is that some inventory is appropriate, if that helps level flow over the entire production process.

Kaizen deployment includes targets and measures for improvement, but these targets, while at times challenging are typically realistic and achievable. While some would argue that if you "shoot to the moon, you might hit the stars," Kaizen focuses on incremental day-to-day success that is sustainable, not a grand one-shot achievement.

A simple suggestion system is also part of Kaizen, and is similar to the process described in the section "Ideas Are Free." Small group activities and teams are also common. Refer to the previous discussion, including the section "Ideas Are Free," and to the discussion of teams and teamwork in Chapter 4.

A final point: process plants, of which Beta has many, already have steady production flow, and don't keep very much inventory in their system (in pipes and tanks). Hence, the need for Kanban becomes substantially less, particularly in light of the risk and consequence of shutting down a particular process. Kaizen principles need considerable adaptation in Beta's process plants.

Beta's Dorton Plant

Beta's Dorton plant was having considerable difficulty. While they had certain structural advantages in lower power costs and ready access to quality raw material that made their costs inherently lower, their operating practices were only average. After some study, the management team decided that applying Lean Manufacturing would help them improve, and that Kaizen methods in particular were key to being lean. So, they worked with a group of consultants who began the Kaizen improvement process with several 5S efforts. These efforts were applied throughout the plant, initially being received with some enthusiasm and sense of pride by the people. They also had some initial improvement in plant performance from the effort, less downtime and better quality. Unfortunately, their efforts were more like 3S than 5S. That is, they did the following:

- Sorting parts and tools and eliminating unnecessary ones.
- Straightening up and put things in order for easy access, using shadow boards for example.
- Scrubbing everything to make it clean, all the while inspecting, detecting, and correcting as appropriate.

Unfortunately, a year or so later, and some six months after the consultants had left, the plant looked much the same as it had before the Kaizen/5S initiative. You could see evidence of their having *been* through a 5S effort—shadow boards, instruction boards, single-point lesson posters, machine/equipment outlines for organizing them, and so on. But, they were no longer using these tools and techniques. Most of the shadow boards had no tools on them; they were scattered about. Equipment was not typically located within the "boxed" areas, but randomly placed throughout. And performance had returned to its level of one year prior.

What went wrong? The improvement method is called 5S for a reason. Dorton failed to do the last 2S's of the methodology:

- Systematize the process to make cleaning and checking routine using the tools developed as part of 5S, and
- Standardize these steps to make them continuous.

The methodology works best as an ongoing effort, and is a waste if it's applied with a "one-shot and we're done" or "check that box" mindset. It's a bit like brushing your teeth; it's something you do everyday, sometimes several times. The plant manager, who was an enthusiast for Kaizen and 5S, had left for a new position at the end of the initial effort. Subsequent to that considerable pressure for cutting costs had come upon his successor, who had little sense of ownership for 5S, or the consultants that came with it. Its death was relatively painless. People went back to their old ways, and it was just another in a series of programs that went by the wayside.

The lessons:

1. 5S must be a way of life, not a program. You may have to initiate some things to get it started, but it must be sustained through the *daily discipline of its application.*

2. Having an improvement process be dependent on the "manager of the moment" is *not* having an improvement process. The process must be aligned to the organization's overall goals and must be a part of its routine way of doing business.

Beta's Burton Plant

Beta's Burton plant was similar to the Dorton plant, but they were much better at keeping up appearances. The plant also had certain structural advantages related to its location and access to raw material that offset average practices in most areas and made the plant appear better than it otherwise would have without these advantages. The plant was quite clean and tidy, at least it appeared so when taking a quick walk-through. However, if you looked closely, for example, you would find: dirt on motors, on and in hydraulic systems and all the equipment; debris behind barriers and in other poorly lit areas; and leaks of all kinds in all the fluid systems nearly everywhere. And, the plant's performance as measured by equipment downtime, quality, and costs, was only marginally better than average. The plant manager seemed more interested in looking good than being good. The plant manager is now coming to grips with these issues, properly applying 5S and making an earnest effort to be good, not just look good.

Summary

Kaizen is focused on continuous improvement and is an integral part of Lean Manufacturing. Key principles include:

- Management must "go to Gemba" (the workplace) and remove the obstacles from worker's success.

- Little innovations and improvements are part of every day's work (which will support "big innovation").

- A strong process orientation is a must for understanding and improving in order to get better results.

- Eliminating waste (any non-value adding cost) in all forms is a must.

- Insisting on high standards and standard work is management's routine responsibility.

- Using 5S, and having excellent housekeeping, is a prerequisite for Total Productive Maintenance.

In the next chapter we'll look in some detail at Total Productive Maintenance.

References

1. Imai, M., *Gemba Kaizen*, New York: McGraw-Hill, 1997.

2. Ford, H., *Today and Tomorrow*, New York: Productivity Press, 2003; original by Doubleday, Page & Company, 1926.

3. Liker, J., *The Toyota Way*, New York: McGraw-Hill, 2004.

4. Robinson, A., and Schroeder, D., *Ideas are Free*, San Francisco: Berrett-Koehler Publishers, 2003.

5. Wheatley, M., *Leadership and the New Science*, San Francisco: Berrett-Koehler Publishers, 1999.

Total Productive Maintenance

9

> TPM is the original equipment-improvement side of the Toyota
> Production System.
>
> *Bob Williamson*

Recall in the previous chapter, that Masaaki Imai[1] clearly states you must first establish the routine use of 5S principles, which will also result in excellent housekeeping, before moving on to total productive maintenance (TPM).

According to Bob Williamson, the name of TPM stems from making the progression from breakdown maintenance to Preventive Maintenance (PM) to Productive Maintenance, and was at one time named Total Preventive Maintenance. As a student and in his discussions with Seiichi Nakajima, the creator of TPM, Williamson relates that Nakajima said in October 1990 that he probably made a mistake calling TPM "Total Productive Maintenance," because it goes well beyond maintenance. Nakajima remarked that the perception of maintenance in the U.S. was fairly dismal, and paraphrasing he said *"You expect equipment to break down. That's why you (Americans) will struggle with the concept of 'Zero Downtime.' You understand Zero Defects and Zero Accidents, but when it comes to equipment you fail to grasp the concept that equipment does not have to fail."* Nakajima went on to say it would have been better to have called it *Total Productive Manufacturing,* since that was its goal. Williamson also observes that his anecdotal information since the late 1980s indicates that the TPM "failure rate" is about 60%. In some cases it may be higher since many so-called TPM efforts were not **true**

TPM, but rather focused on the single element of having operators do autonomous maintenance. One of the biggest causes of failures in TPM implementation is organizational leaders not fully understanding TPM, and thinking it to be another "maintenance program," and delegating its leadership to the maintenance department. It is wrong to do this, and nothing could be further from the truth.[2]

Fundamentals of TPM

So, what is TPM, or if you prefer as Nakajima later admonished us to consider, Total Productive Manufacturing? Below is a summary of the key elements of TPM according to Nakajima.[2,3]

- *In a TPM culture, maintenance is about maintaining the plant and equipment function, that is, **prevent its breaking**;* unfortunately, in U.S./Western cultures, maintenance is *too often* about repairing equipment, that is, "fixing stuff" after it breaks; we expect equipment to break down. This represents a huge psychological difference, and potential competitive advantage between in a typical U.S./Western plant and one with a TPM mindset. A more appropriate name for TPM is Total Productive Manufacturing, maintaining the plant's function!

- In a TPM culture when equipment is new, it is as bad as it will ever be; or, said differently, the current state of our equipment is always its worst state, relative to its future; we're going to constantly improve it. Unfortunately, in most U.S./Western cultures, when equipment is new, it is as good as it will ever be; the attitude seems to be that it's just going to get worse with time.

- TPM calls for measuring overall equipment effectiveness and all losses from ideal production capability (24 hours per day, 7 days per week), i.e.,:

  Overall Equipment Effectiveness (OEE) = Availability% × Rate% × Quality%. For example, if A = 95%, R = 95%, Q = 95%, then:

  OEE = A × R × Q = 86%

- These losses from ideal can be viewed as literal or potential waste. We'll examine the details of OEE and discuss ways to measure and minimize this waste later in this chapter.

- TPM calls for restoring equipment performance to a *like-new condition, or better*, not simply "patching" it to quickly get it back on line, something that occurs all too often in U.S./Western plants. Clearly, restoring equipment to like-new, *or better*, is consistent with the concept of continuous improvement. Not doing so is its antithesis.

- TPM calls for operator care and involvement in maintaining equipment, and particularly its functional capability. Some call this autonomous maintenance. Here, operators take care of the equipment (e.g., tighten, lubricate, clean, and work to avoid defects and failure modes, or as appropriate to monitor them, so that their consequence can be minimized).

- TPM calls for improving maintenance efficiency and effectiveness, that is, you must understand your defects or failure modes, and work proactively to avoid them, and/or to detect them early enough to minimize their consequence.

- TPM calls for training people to improve their job skills. Increased variability in process and equipment performance is often induced by poor training and skills in the application of standard work. Having better skills helps reduce the variability, and helps to eliminate the defects, failures, and waste that would otherwise occur.

- TPM calls for new equipment management and maintenance prevention *in the design*. Once we understand the nature of the defects and failure modes, we work to stop these through better design of the equipment and of the work practices.

- TPM calls for the effective use of planned, preventive, and predictive maintenance technology. PM includes routine tender loving care (TLC), such as tightening, lubricating, cleaning (cleaning includes inspecting for defects), and periodically taking care of the equipment. Predictive maintenance (PdM) is knowing the condition of your equipment so you can better take care of it, and minimize the consequence of an impending failure. We'll review predictive maintenance more in a later chapter. Planning is anticipating a requirement, organizing it in adequate detail, scheduling it, and in effect managing it, to minimize the waste and cost, and to maximize the benefit.

Let's consider some of these in more detail.

OEE

The purpose of measuring OEE is to understand your process/plant's losses from the ideal. These losses are waste that must be minimized. Knowing all losses from the ideal helps prioritize resources for action to minimize the waste in your system. Examples of loss categories include:

- Scheduled/planned downtime (e.g., through maintenance and production planning)
- Unscheduled/unplanned downtime (e.g., for equipment failures and spurious events)
- Rate losses (e.g., reduced running rate, short stops, and change-overs)
- Quality losses (e.g., rejects and rework)

The differences between maintenance planning and scheduling are often confused. A discussion of this topic is provided in Appendix A.

Some of Beta's plants have created categories of losses that are specific to their business, for example, raw material quantity and quality might be a specific loss category not listed here that is important for a given plant; or utility supply losses due to inadequate power, water, compressed air, and so on, might be a specific category. Also, the level of detail for each will vary depending on the specific business requirements. Many businesses will also differentiate between production losses: those controlled by the operating units, and those due to poor marketing and sales and general market and economic conditions.

A question that most manufacturing plant managers should be asking is:

If you could run your plant 8,760 hours per year, making 100% first-pass, first-quality products, at 100% of your maximum demonstrated, sustainable rate, with no losses for product change-overs, how much could you make? How much are you making? Where are your losses from ideal? Most importantly, *are these losses acceptable to your business?*

In other words, what's ideal, and how close are you to that state. Some of these losses may be acceptable. For example, if your unplanned downtime is <1%, and you simultaneously believe that: (1) other losses are much easier to address and have much more value, and (2) with current technology achieving much <1% would be cost prohibitive. If so, then your current performance might be acceptable, *for now*. Or, for example, suppose current market demand is such that you only need to run two shifts to meet demand, then production losses due to a lack of demand might be acceptable, *for now*. You would still continue to measure these and make business decisions about each one, so that you minimize the losses and waste to your business.

OEE is a general indication, and not necessarily a goal. Quantifying and managing actual losses and making business decisions about each loss are key to improvement. Further, while the measure is managed by the plant management team, the day-to-day ownership of these losses should be driven down to the shop floor where direct decisions can be made about reducing them, and management must facilitate their success in doing so, removing any related obstacles. Figure 9-1 provides a model for measuring OEE, and minimizing losses from ideal or waste.

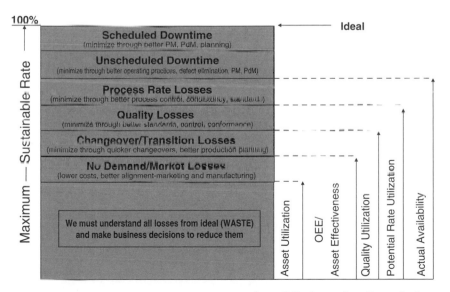

Source: *Making Common Sense Common Practice*,
Elsevier Butterworth-Heinemann Books,
Boston, MA, & London, England

Figure 9-1 *OEE/AU and Loss Accounting Model—Minimize Waste.*

If each category is considered a waste, then how do we reduce these losses from ideal? Below are suggestions for minimizing the waste in each category:

Scheduled downtime—reduce this waste by improving:

- Production and maintenance planning and scheduling
- Shutdown and overhaul practices
- Equipment reliability and life
- Design and operational practices for operability and maintainability
- Minimization of any unnecessary PM

Unscheduled downtime—reduce this waste by:

- Improving equipment reliability through better design, operation, and maintenance practices
- Applying TPM and reliability-centered maintenance (RCM) principles for operator care, PM, and eliminating failure modes
- Applying condition monitoring to detect onset of failure early enough to plan and minimize the consequence of the pending failure
- Better installation, startup, and commissioning of equipment and processes (note that as much as 67% of failures occur in the infant mortality period)

Process rate losses—reduce this waste by:

- Improving process conformance through:
  - equipment reliability
  - proper, calibrated instrumentation
  - disciplined operation using standard operating procedures and conditions
  - quality and timely raw material
- Operational discipline in doing standard work

- Consistency of process between shifts (i.e., common set points and standard practices across all shifts to minimize variability)
- Good shift handover practices to assure teamwork and consistency across all shifts

Quality losses—reduce this waste by:

- Getting the basics right
- Disciplined use of simple statistical process control (SPC) (trend and control charts)
- Measurement of first-pass, first-quality yield, and reasons for failure to achieve this
- Application of Six Sigma principles, as appropriate
- Discipline in all other loss minimization

Changeover and transition losses—reduce this waste by:

- Quick changeover practices (e.g., set up reduction and "Single Minute Exchange of Die" [SMED]) (see discussion below)
- Improved production planning
- Improved equipment/process reliability
- Rationalization of product mix and customers by applying supply chain management principles

Market losses—reduce this waste by:

- Striving to reduce costs: innovation on the shop floor leading to more competitive cost position and higher gross profits for innovation in R&D and new process and product development
- Being innovative in product and process development
- Improved market positioning by understanding "what wins orders?"
  - Price (and costs that allow adequate gross profit)

- Delivery
- Quality
- Unique features or capability
- Technical support
- Aligning your manufacturing and marketing strategies

Suggestions are provided in Chapter 11 on Supply Chain Management for aligning the marketing and manufacturing strategies.

OEE/TPM and Safety Performance

Several studies indicate that improved OEE and safety go hand in hand. Figure 9-2 illustrates this point well. The data shown here, which is from Beta's Tomahawk division, is only one of several sets of similar data. The data has been "normalized" to a fixed number for both OEE and injury rate and, thus, reflects percentages of those numbers. The data is compelling. If safe behavior is a requirement at your plant, then best practice, as you define it (that yield better performance), should be a requirement as well. Fewer interventions result in fewer injuries and less risk. Thus, practices should assure that we avoid the failures that result in the interventions. All this demands that we have disciplined, standardized practices in all that we do. And, that we constantly look for ways to do these even better.[5]

TPM calls for excellence in preventive and predictive maintenance as a means of improving OEE and reducing waste. This excellence also results in fewer injuries as illustrated in Figure 9-3, which is from Beta's Brainard plant's work order history over an eight-year period. At that same plant, related data shown in Figure 9-4 demonstrates that an increase in reactive work also tends to result in more injuries.

The message is clear: get your practices right and your overall performance in all areas will improve. And as we saw in Chapter 2, Beta has observed that we can use the same steps used in a previous safety initiative to drive manufacturing and operational improvement[6]:

- Top-down leadership—clear, consistent expectations
- Bottom-up ownership and employee engagement

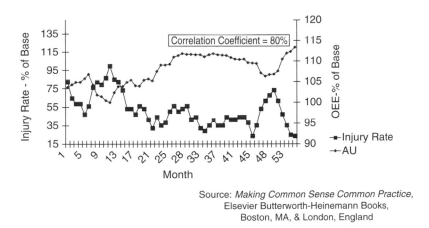

Source: *Making Common Sense Common Practice*,
Elsevier Butterworth-Heinemann Books,
Boston, MA, & London, England

Figure 9-2 *Injury Rate vs. OEE With Time.*

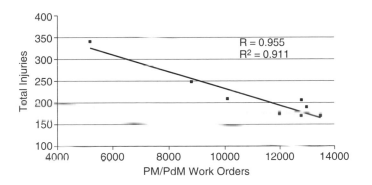

Figure 9-3 *Correlation of PM/PdM Work Orders with Injury Rate—Plant #1.*

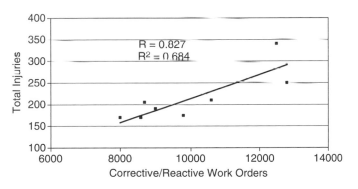

Figure 9-4 *Correlation of Corrective/Reactive Work Orders with Injury Rate—Plant #1.*

- Education and training
- Action plans and measures
- Visual communication
- Standards and procedures
- Benchmarking and aggressive goals
- Audits and assessments
- Root cause focus—eliminate repeat failures
- Rewards (and punishment)
- Resources for supporting improvement
- Continuous improvement expectation and process
- A culture....a way of life

If you believe in zero incidents/injuries as an uncompromising goal, then you must believe in zero failures/downtime is an uncompromising goal. Manufacturing excellence and safety go hand in hand. The leaders of any organization have a clear responsibility to assure both, applying the same principles.

Operator Care

While operator care and so-called autonomous maintenance are critical elements in TPM, it's essential that we understand that it is only an element, and not the entirety of TPM, as some appear to believe. That said, it's difficult to overstate the importance of TLC through efforts like Tightening, Lubricating, and Cleaning. Dirt is the enemy of reliability. As John Cray said, "We clean to inspect, we inspect to detect, and we detect to correct." It's a simple saying with powerful implications. It's also essential that we understand TPM should more likely be called Total Productive Manufacturing, where the philosophy is to maintain the plant's function; to minimize the waste in the manufacturing process; and to continuously look for ways to improve the manufacturing process, not just maintenance.

With this in mind, let's consider a model for operator autonomous maintenance, as suggested by Ian Gordon. His description of the duties of what he calls an operator/owner and a maintainer/improver are provided below.[7]

Operator/Owner Guidelines

The operator/owner must do the following:

- Observe equipment performance: variations in process and abnormalities in condition, and work to minimize these and manage them.

- Operate within agreed process limits and measure any variations outside those limits, using the measures of these variations to prioritize issues for improvement.

- Maintain good equipment condition (e.g., tighten, lubricate, clean, inspect, and manage environmental factors).

- Do daily monitoring and inspection: look, listen, feel, smell; eliminate obvious defects; monitor process condition for conformity to requirements. (Clean to inspect, inspect to detect, and detect to correct.)

- Identify deficiencies in the equipment operation and design for changes thereto.

- Prioritize improvement work for the support team and related resources.

- Measure/estimate OEE and categorize losses from ideal at each shift. Use these losses to prioritize improvement activities.

- Work with maintainer/improver to balance preventive maintenance, predictive maintenance, operator care activities, and to better plan and schedule both maintenance and operations.

Maintainer/Improver Guidelines

The maintainer/improver supports and works with the operator/owner to make things better. This is not to say that the operator/owner has no responsibility for improving. Clearly they do, but rather it is to add emphasis to the fact that maintenance is part of the improvement process too. The maintainer/improver has the following responsibilities:

- They are the technical expert in the appropriate field, for example:
  - Mechanical
  - Process control/instrumentation/electronics
  - Electrical/power
- They focus on defect elimination, that is, they analyze and measure equipment performance and potential defects (in addition to the operator doing so); and support root cause analysis, again along with the operator.
- They support the operator/owner regarding:
  - Standards for inspections, lubrication, minor adjustments, or routine care/service
  - Assistance in diagnosis of problems and variations in performance
- They lead the development of the PM plan, for example:
  - Time- and condition-based procedures and results
  - Timing and scope of corrective work needed
  - Oversee large pre-planned maintenance work
  - Evaluation of the effectiveness of PM/corrective work
- They lead the development of the proactive maintenance efforts, e.g.,:
  - Support of root cause analysis and design-out efforts
  - Participate in design reviews for new projects
  - Support commissioning of new/modified equipment
- They provide systems support, for example:
  - Review/upgrade maintenance procedures, specs, and tasks to assure reliability
  - Assure excellence in spares and bills of material
  - Assist with technical documents and drawings
  - Maintain equipment performance and maintenance history

- Contribute to the reliability improvement plan
- Assist in asset management planning
- Mentor junior engineers
- Beta is using this model at several of its plants.

Quick Changeover

Quick changeovers facilitate the reduction of lost time during a product change. The principles of quick changeover resulted originally from the concept of SMED, that is, changing a die in a single minute, something that at the time was considered unrealistic. Bob Hansen offers a simple set of steps for improving changeover practices and reducing the waste that often results. The steps he suggests are as follows[4]:

1. Assemble a cross-functional changeover improvement team.
2. Videotape a typical changeover event.
   a. You may need several synchronized cameras to fully capture all the activities.
   b. Include in your video any appropriate preparation and follow-up actions.
3. Review the video as a team, and document the following:
   a. Each action taken during the changeover
   b. Its elapsed time, and
   c. The reason for each action.
4. Analyze each action: define whether it is *internal* to the process, or *external*.
   a. If internal, it must be done within the changeover process and timeline.
   b. If external, it could be done offline in preparation for the changeover effort.
5. Review all internal efforts to determine if they can be performed in parallel.
6. If they cannot be performed in parallel, analyze the internal critical path items for those that can be reduced in time (e.g., using quick-lock connections vs. screws, redesigning fixtures

and dyes to mount to pre-fitted guideposts or points, or doing preheating or other such activities outside the changeover).

7. Review resources availability and skills for reducing the time-line—people, parts, and materials. Are additional resources justified?

8. Plan the revised changeover process, and train all involved in the proposed changes.

9. Do the changeover in the new format two to four times.

10. Review the new changeover for additional improvements, and modify accordingly.

11. Train everyone in the methodology and apply it in a disciplined way.

Relationship Between TPM and RCM

A question sometimes asked is whether or not TPM and RCM are compatible. Since RCM will be discussed in a later chapter, we will review the relationship between RCM and TPM there.

Beta's Ivyton Plant

Ivyton is one of Beta's chemical operations, and like all Beta's other operations it is under intense pressure to improve. After reviewing various improvement tools, the division manager of this operation concluded that TPM would be the appropriate tool to use, and authorized beginning an effort to apply those principles. He reasoned that maintenance costs were a large portion of their operating costs, typically running 20% to 25%, and unplanned downtime was inducing large production losses and other costs, all of which combined to put the entire operation at risk. Having a more productive maintenance function, the division manager reasoned, made a lot of sense, and so he picked two plants to "pilot" a TPM implementation effort. Note, however, that this manager is taking the phrase literally, that is, TPM is a way of making maintenance totally productive, and in particular, of getting operators to do maintenance. While operator care is a good thing to be done, and not withstanding the initial review, clearly this manager does not understand the basics of TPM. One of the plants selected was the Ivyton plant. A review of its performance and tour of the site yielded the following:

1. Housekeeping was awful, and the application of 5S principles was essentially non-existent. Recall that this is a prerequisite for TPM.

2. The plant was relatively old, and most felt had outlived its functional life. The attitude was not one of making it better, but rather of just keeping it going.

3. Maintenance was viewed as a repair function, not a process for maintaining the plant's function. Restoring equipment to a like-new condition, or better, was not a requirement for the maintenance that was done. Most maintenance was done with a focus on getting the plant back on line as quickly as possible, not as correctly as possible.

4. OEE was not being measured. Tons were, and unit costs were, but OEE and losses from ideal performance were not. Thus, they had no way of really knowing what the priorities should be relative to production losses. Most of the analysis, if you could call it that, was anecdotal.

5. Operator care was not being done in any recognizable form. Indeed, the union at the plant was very much opposed to the concept of operator care. They thought of this as operators doing maintenance, which would result in maintenance technician layoffs. This was a big impediment to the division manager's goal for autonomous maintenance.

6. PM, PdM, and planned maintenance were very poor. Most maintenance was done on a reactive basis, and very little effort went into doing PM, PdM, or planned maintenance.

7. Training in skill development was limited to about 20 hours per year per person. Even if money were available for training, it was difficult to get the people into a classroom for training, given all the emergency ongoing work.

The result was a plant that was in disarray, one where maintenance was being blamed for most of the plant's problems. Clearly, opportunity existed, but a fundamental shift had to occur regarding the meaning of TPM. The first step taken was to educate the plant and division manager about the meaning of TPM. Immediately the name of the process was changed to Total Productive Manufacturing. Additional emphasis was placed on all the elements of TPM, but especially on the prerequisites of housekeeping and the 5S principles to prepare for TPM. Other steps were taken to implement an OEE measure, to establish a simple operator care program consistent with the union contract, and to do a much better job with planned, preventive, and predictive maintenance. Today, Ivyton is well on its way to being a much better plant, thanks to much better leadership and specifically to their understanding that TPM is more than doing operator maintenance (Total Productive Manufacturing). Ivyton now has a good process for managing change and much stronger plant leadership. The union is now more engaged and has less fear of the TPM model as a job cutter, and is participating in TPM improvement efforts. The plant has a long journey ahead, but the key elements of the change process are in place.

Beta's Grethel Plant

At Beta's Grethel plant, a process-type operation, the situation was essentially opposite, except for the pressure to improve. The use of TPM principles began from the initial operation of the plant, and the plant manager was fully supportive and had an excellent understanding of the TPM principles, along the lines of Total Productive Manufacturing, not just better maintenance. A review of its performance and a tour of the plant showed the following:

1. Housekeeping was excellent, and the routine, disciplined application of the 5S principles was evident. Recall that this is a prerequisite for TPM.

2. The plant was relatively new, and had done TPM from the beginning of its operation. The attitude among employees was one of continuously making it better, so it would be there for decades to come.

3. Maintenance was viewed as a support function to production, which considered maintenance to be a process for maintain-

ing the plant's function; one related requirement was to restore equipment to a like-new condition, *or better*.

4. OEE was being measured in detail, with the ownership of the losses from ideal typically being with production. Lost profit opportunity for each percent of OEE and each unit of lost production were well understood. OEE and particularly minimizing the losses from ideal was used as a driver for performance improvement.

5. Operator care was being done regularly to a reasonable extent. Note that Grethel is a process plant where operator care is more difficult because of the size of the area for which each operator was responsible. However, the operators and maintainers had worked together to see what operators could reasonably do to assure basic care of the plant.

6. PM, PdM, and planned maintenance was excellent. There was considerable PM, operator care, and predictive maintenance being done, which was flowing into, or from, an excellent work management system for planning and scheduling, and getting PM done.

7. Training was encouraged and supported, with a minimum target of 40 hours per year in said employee's skill development.

The result was a plant that was in excellent condition, with high production and low unit costs. And yet, they felt that they had abundant opportunity for improvement. They truly understood the TPM principles as being focused on Total Productive Manufacturing, and were applying them. Grethel continues to work hard to improve its performance. It has a bright future.

Summary

Beta's experience is that while TPM *principles* apply to all plants, as a practical matter the actual practice of TPM is better suited to its batch and discrete plants, where operators have the equipment "at hand" and can play a greater hands-on role in equipment care (TLC) while measuring losses from ideal production. TPM can be more difficult at large process plants where the equipment is spread over a much larger area, making application of the principles more difficult. On the other hand, the planning part of TPM seems to have greater importance at process plants, given the complexity, risk, and performing maintenance efforts there. In any event, considerable adaptation of TPM is required in process plants.

The OEE measurement applies equally well in both batch/discrete and process plants, and provides good guidance as to which problem represents the greatest opportunity for improvement and the elimination of potential waste. As Bob Williamson has said "OEE loss elimination guides everything in TPM."

Training is essential for the reduction of defects and the elimination of variation in your production processes. Training and skill development are paramount in your business. As the old saying goes, *if you think education is expensive, try ignorance.*

Operator care is of vital importance, much the same as we care for our cars, operators must care for the equipment. Besides, in Beta's operations there are typically 3 to 10 times as many operators than maintenance technicians. Thus, it's essential to have operators care for the equipment, adapted of course to each operating circumstance. As Charles Bailey, Vice President of Operations of Eastman Chemicals observed, "Reliability cannot be driven by the maintenance organization. It must be driven by the operating units... and led from the top." I would add that to do otherwise is a bit like expecting the mechanic at the garage to "own" the reliability of our cars. They can't and won't. We must own the reliability of our cars, asking the mechanic (maintenance) to help us. However, the analogy is not quite this simple; make no mistake, the maintenance staff are an integral part of the operating units and must support achieving reliability and manufacturing excellence, but they alone cannot deliver reliability.

Having excellence in planned, preventive (PM), and predictive maintenance (PdM) are essential. PM and PdM are planned and scheduled, and the results of these efforts identify the corrective work that must be planned and scheduled for maximum effectiveness and efficiency of the work, and to minimize its adverse consequence to the business.

A strong focus on *maintenance prevention* is critical in the continuous improvement model; changing the design, operating, and maintenance practices so that we do less maintenance is essential. Stopping the defects which result in the maintenance requirement and maintaining equipment function is paramount. Properly applied, all of these methods will help in that endeavor.

References

1. Imai, M., *Gemba Kaizen*, New York: McGraw-Hill, 1997.
2. Nakajima, S., *Total Productive Maintenance*, Portland, OR: Productivity Press, 1993.
3. Williamson, R. W., Personal Communications, February, 2006.
4. Hansen, R., *Overall Equipment Effectiveness*, New York: Industrial Press, 2001.
5. Moore, R. *Making Common Sense Common Practice: Models for Manufacturing Excellence*, Boston, MA and London, England: Elsevier Butterworth-Heinemann, 2004.
6. Leonard, J., Personal Communications, 2005.
7. Gordon, I., Personal Communications, 2001–2005.

Six Sigma

You shouldn't attempt to do statistically driven improvements until
you have a steady process.

W. Edwards Deming

Jack Welch was enormously successful at General Electric. His success
is, at least in part, attributed to the use of Six Sigma in improving
their operational and resulting financial performance. If we use Six
Sigma, the logic goes, then our company is more likely to be successful. While this logic is admittedly too simplistic, it does seem to ring
a cord in many circles, perhaps more so in companies looking for the
next big step change in improvement. However, it may be that step
change or overnight success takes about 10 years, no matter what
course you take. Below is a review of Six Sigma according to Pande
et al.,[1] along with additional commentary.

Definition of Six Sigma

Taken literally, Six Sigma is a statistical term which characterizes your
quality having <3.4 defects per million for a given product or process
specification. Table 10-1 provides the defect rate for one through six
sigma levels of performance.

However, Six Sigma has become a methodology for reducing the
variability of processes such that the result is greater quality and con-

Table 10-1   *Defect Rate Per Million: Various "Sigma" Levels*

Meets Specification Rate	Defect Rate	Sigma
30.9%	691,000 ppm	1
69.2%	308,000 ppm	2
93.3%	66,000 ppm	3
99.4%	6,210 ppm	4
99.98%	320 ppm	5
99.9997%	3 ppm	6

sistency and better performance. According to Pande et al., it stresses the simultaneous achievement of seeming contrary objectives:

Being stable, and innovative

Seeing the big picture, and the details

Being creative, and rational

Let's consider each of these.

Being stable and innovative. This concept is very similar to the Lean Manufacturing concept of standardization and innovation being two sides of the same coin. It's essential to have stability before you can improve. Without stability, it's hard to determine where you are, and whether or not you have improved. It's a bit like hitting a moving target.

See the big picture, and the details. This is also similar to Lean Manufacturing, which requires that we have long-term thinking, and that we look at our manufacturing process as a value stream from design through to service and support and to minimize the cost of the system. Six Sigma requires that we see the "big picture" and understand the impact a decision in one part of the system will have on other parts of the system. And like Kaizen, it also requires that we get the details right, making the hundreds of day-to-day improvements that will assure excellence.

Being creative, and rational. Again this is similar to Lean Manufacturing, which requires that we constantly look for the small innovations (the rational continuous improvement side) so that we can create an environment where we constantly look for and can finance the big creative steps in our products and processes that represent a step change in improvement.

The Methodology

Pande et al. indicate that a focus on its themes of genuine customer satisfaction/success and feedback, data and fact driven management, process management and improvement, proactive management and continuous improvement, boundaryless collaboration, and a simultaneous drive for perfection and tolerance for failure are key factors to its application. Included in the methodology is understanding and improving your business processes, using statistical process control (SPC) techniques and balanced score cards, and using improvement projects to capture the value identified in the analyses.

These too are similar to Lean Manufacturing. Customer satisfaction in quality, cost, and delivery is the first priority of business in Lean Manufacturing. Indeed the value-add analysis (and the elimination of waste) begin with the customer. Value stream mapping, or understanding your business processes, is used to identify those areas where waste, or non-value adding costs, are incurred in the system. Data and facts from various sources are used to make decisions, (e.g., the people doing the work and the defects they see, the Overall Equipment Effectiveness [OEE] measure and losses from ideal, the waste observed in day-to-day efforts, the non-conformances that result from SPC measurements). Lean Manufacturing also requires the use of balanced measures to reflect system performance, the use of andon to signal a performance measure that is not in control. It supports individuals and small teams quickly implementing solutions with little or no red tape, but then standardizing the solution so that everyone will consistently follow it (and minimize variation). It calls for routine collaboration among team members (improvement time) in cross-functional teams to develop the best solution. Finally, Lean Manufacturing strives for perfection, but considers a failure an opportunity for learning how to do the job better the next time.

Similar to Deming's "plan, do, check, act," Pande et al.'s Six Sigma approach applies the DMAIC model (Define, Measure, Analyze, Improve, and Control core processes and key customers). Principal tools/methods include:

Process design/re-design

Process management

Continuous improvement

Customer feedback

Creative thinking

Analysis of variance

Balanced scorecards

Design of experiments

Statistical process control

Improvement projects

Finally, Pande et al.'s Six Sigma approach applies the following roadmap:

1. Identify core processes and key customers

2. Define customer requirements

3. Measure current performance

4. Prioritize, analyze, and implement improvements

5. Expand and integrate the Six Sigma system

Again, all these elements are generally included in the Lean Manufacturing model. So, if they're so similar, why are they sometimes perceived differently?

It appears to be related to the actual application of these principles. Toyota, for example, works hard to keep things simple, encouraging people on the shop floor to make decisions without a lengthy analysis or, for example, without a project that is authorized based on a statistical analysis. They don't have "green belts" or "black belts" as are common in Six Sigma. They do have considerable training in processes and standard work, but appear to let the folks doing the work make the decisions based on "common sense," and their experience with the process. Kaizen and Lean Manufacturing do use the "plan, do check, act" methodology, as well as its parallel methodology once the standards have been agreed—"standardize, do, check, act." What appears to be different is that the DMAIC methodology can require extensive data collection, measurements, and analysis so that one can clearly define the nature of the problem, so that a solution can be implemented and controlled. While it does provide defensible data for the improvement project proposed, it also appears to take more time to develop the data and do the analysis before a solution is implemented.

An Alternative Application of Six Sigma Principles

Figure 10-1 provides an alternative way of thinking about Six Sigma and achieving it.[2] David Burns developed the concepts shown and he observes that *most* plants operate somewhere near Three Sigma, that is about 93% of the time their processes are in control and are meeting specifications or requirements. He also observes that a competitive advantage is created when a given business achieves Four Sigma, wherein more than 99% of their processes are in control and meeting requirements. The way to achieve that is to fix the obvious, and do the basics really well by fully engaging the entire workforce in the continuous process improvement. His experience and data support this view, and it is entirely consistent with Toyota's performance in particular, and Lean Manufacturing in general. In Burns' model, the first step in applying Six Sigma is to fix the obvious and

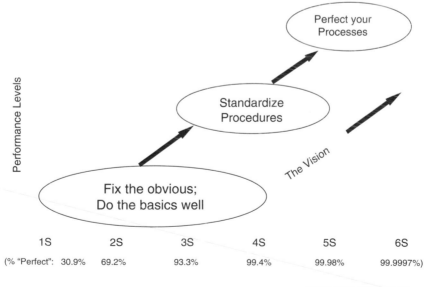

Source: David Burns, Reliability
Services Ltd. Melbourne, Australia

Figure 10-1   *Six Sigma: Focus on Minimizing Variability.*

make sure good basic practices are in place and practiced in a disciplined, consistent manner; the second step is to standardize those processes; and finally step three is to seek perfection in the processes. Of course applying the tools previously described for doing this is a good approach. The point of emphasis here, however, is that achieving step one—getting good basic practices in place first, including excellence in maintenance and reliability, is essential for success. One other observation is that if you are ever able to actually perfect your processes, they're likely to change, resulting in a never-ending search for perfection.

Supporting Burns' view, Winston Ledet says, "We have run some very interesting scenarios in our dynamic benchmarking model that shows how getting all of those other small things done is a great *precursor* to doing Six Sigma. I guess it is like Deming said, you shouldn't attempt to do statistically driven improvements until you have a steady process."[3]

Beta's Waldo Plant

Beta's Waldo plant had been working to improve its performance and was having considerable success. Waldo had modeled its entire production process using a statistical modeling tool, and had input reliability data for each process based on historical data for availability, rate, and quality performance, and could predict with considerable accuracy and confidence its ability to produce the required product at a given level during a given period. This is not to say that it was producing at the level desired nor at the cost desired. But, Waldo did understand its current performance and capability.

However, Waldo wanted better results, and so it began the implementation of a Six Sigma program, as part of an overall corporate Six Sigma program. People were trained in Six Sigma, resulting in many "green belts" and several "black belts." The black belts in turn were required to identify projects that they would pursue that would result in substantial improvement or savings, applying the DMAIC process, and they did so. After about one year of effort, however, the Six Sigma program seemed stalled. About a million dollars had been invested into the program, but little savings or improvement had resulted.

Because of this, three major Six Sigma projects were reviewed with a view to see if and/or why the program was stalled. During one of the reviews, the "auditor" stopped the review, commenting to the "black belt" engineer that he did not want to hear any more. Surprised, the "black belt" asked why? The auditor replied that the project for which you are collecting and analyzing data has not yet started, but it does have a known solution that has a very high probability of being effective and saving money, if properly implemented. He was advised to get on with it. He balked asking if we didn't need to collect more data and do more analysis. The auditor responded that it wouldn't improve the probability of his success. And, that was the auditor's advice for all the other projects at Waldo that were using the Six Sigma improvement tool. Unfortunately, they were focused much more on getting the process right than they were on getting results. This resulted in a condition sometimes referred to as 'paralysis by analysis,' and less than desirable results.

Beta's Staffordville Plant

Beta's Staffordville plant was generally good in its manufacturing practices, but was excellent when measured by customer returns, achieving better than a Six Sigma level of performance, that is, <3 returns per million units of product delivered. At the same time, its unit costs were quite high relative to other plants and its competition. Part of the reason for this was that Staffordville had an excellent inspection program for its finished product, and each unit had to go through a final inspection that was extensive and expensive. While they were achieving Six Sigma in product quality delivered to the customer, it was coming at a price—lots of scrap and reworked product. Staffordville needed to push the quality practices back into each of the processes and make sure that each step of the production process did not pass along defects to the next process, applying Lean Manufacturing principles.

Beta's Teaberry Plant

Beta's Teaberry Plant was doing reasonably well, but was concerned about its future and the impact of Asian competition. Six Sigma methods were introduced at Teaberry with initial goals including reducing waste/scrap, reducing material costs through greater process consistency, and improving overall productivity.

Project teams were formed in each area and were initially led by an external Six Sigma facilitator, who served as trainer and coach. The teams used the DMAIC methodology in each area, with a focus on production process consistency. The teams initially defined the key process variables, they then measured current performance of those variables, identifying those areas and issues that if improved would lead to greater consistency. They analyzed their performance and those key influencers and subsequently put forth an improvement plan that included better instrumentation and calibration standards, improved procedures and checklists, training in those improved procedures, better shift handover for assuring consistency across shifts. Over time they monitored and measured the results of their effort.

The result: 10% lower material and waste/scrap costs (note that material costs were 50%+ of their overall costs, so this was huge), 25% less overtime (mostly for sorting out and/or recycling the scrap/ waste), and a 12% increase in overall throughput. The plant was becoming much more stable and, thus, had greater production capability. And, profitability was improving dramatically.

Six Sigma and Lean Manufacturing

There are some who support the combination of Lean Manufacturing and Six Sigma, and indeed many similarities are discussed above. However, there are others who are well experienced in lean manufacturing who do not believe that Six Sigma and Lean Manufacturing can be combined. Jeff Liker said "I do not believe lean tools or Six Sigma tools or a marriage of the two will get a company to a lean learning enterprise."[4] Gary Convis, President of Toyota North America, said that it took him 10 years to understand the Toyota Production System (TPS), and he is still learning today. He goes on to observe that there are companies seeking to benefit from TPS or Six Sigma who send employees to train for two weeks, and ask them to do a project, and then anoint them as experts.[4] The implication of his comment being that this is not realistic. If one interprets Six Sigma as doing the basics really well and standardizing work, then continuing to improve it, then Lean Manufacturing and Six Sigma are quite compatible. On the other hand, if Six Sigma is applied like it was at Beta's Waldo plant, then it will be very difficult to make it a cost-effective tool. From an engineering perspective, Six Sigma is a great tool; it's thorough, structured, data-driven, fact-based, and so on. However, for the business we should be more circumspect. It may be more appropriate to get the basics right first, getting our systems stable and reliable, and hold Six Sigma for the more difficult problems, ones that require a comprehensive problem-solving approach, and not "blanket" a plant or business with a rigorous Six Sigma process.

Summary

Six Sigma is about reducing the variability of your processes. Lots of variability is introduced from simply not following procedures and having standard work, or not doing the basic practices well. Think of doing the basics as "*little*" Six Sigma. Recall Figure 10-1 that states that to get to Four Sigma (99.4% spot-on), you must fix the obvious and get the basics right all the time in order to achieve a competitive advantage. This is my advice to those who would venture into the Six Sigma world. If you get too engrossed in the process, you may forget that your goal is to get results. The result of this may be a highly refined process that doesn't deliver any improvement. Further, you don't need Six Sigma to tell you that you must follow your standard operating procedures rigorously across all shifts to minimize the variability of your processes, nor do you need Six Sigma to know that you must have excellence in your lubricating practices—right lubricant, right ISO cleanliness standards, right frequency, and right quantities; nor to know that you must properly align and balance your machinery. These are things that don't need analysis. They just need to be done, all the while looking for simple ways to make them even better. Six Sigma should be reserved for the more complex problems that do not have known or obvious solutions, and it should focus on getting results at least as much as it does the analysis.

Six Sigma is a great tool. But, it's a bit like RCM, as we'll see in the next chapter, in that it consumes considerable resources in applying a relatively complex improvement process, and thus *people often get so engrossed in the process they forget the goal is to get results.* It should be selectively applied to complex problems that require a disciplined methodology. Otherwise, you can get tied up in "paralysis by analysis" and not get significant results, as Beta's Waldo plant seems to have demonstrated. General Electric appears to have been successful with Six Sigma at least in part because of the drive of Jack Welch and his demand for results. It's also reported that he felt that it was a good tool, but that other tools may have sufficed. We should typically be using Six Sigma to narrow the focus on problems that can provide results within three to six months, which is about the attention span of most people before they get frustrated with any process. The business-level Failure Modes And Effects Analysis (FMEA) can also be used to narrow the field of potential projects to which Six Sigma might be applied. Some additional observations and suggestions are provided below:

- Don't start your Six Sigma effort with a "nerdy" monologue on SPC or statistical analysis—this damages its credibility, particularly with the shop floor.

- General Electric has been very successful with Six Sigma because the leadership drove the process and demanded results.

- Toyota has been very successful without Six Sigma, using simple tools, Kaizen, 5S, TPM, and engaging the entire workforce in improvement.

- Which one is best? Start with the Toyota model and do the basics really well first.

- Lots of variability is introduced in:
  - Unreliable equipment because of dirty oil, poor operation, poor maintenance
  - Unstable processes
  - Lack of standard work (e.g., shift-to-shift variation)
  - Poor operator care/TLC
  - Poor process conformance
  - Similar kinds of activities

- Get basics of operating and maintenance right first—simple defect elimination

- Use the business-level FMEA as well as OEE/AU and losses from ideal, and equipment histories for selecting improvement projects.

- Reserve Six Sigma for the difficult problems, and/or for supporting your business case/funding.

Beta plans to follow these guidelines, getting the basics right first, applying Kaizen and TPM, and reserving Six Sigma for the more complex problems.

References

1. Pande, P., Neuman, R., and Cavanaugh, R., *The Six Sigma Way*, New York: McGraw-Hill, 2000.

2. Burns, D., Personal Communications, 2001–2005.

3. Ledet, W., Sr., Personal Communication, 2004.

4. Liker, J., *The Toyota Way*, New York: McGraw-Hill, 2004.

Supply Chain Management

11

The quality of our offerings will be the sum total of the excellence
of each player in our supply team.

Robert W. Gavin

Supply Chain Management: One Model

In a presentation at the University of Dayton and in his book *Beyond Negotiation*, Robert Walker offers a methodology for applying supply chain management to improve company, supplier, and customer performance. In his view, a supply *chain* is a supplier, a manufacturer, and a customer, linked together and working as a team to optimize chain performance. Below, a summary of his model for managing a supply chain[1,2] is discussed.

Applying supply chain principles can minimize business system cycle time, inventories, and costs, while assuring a timely delivery, to improve: (1) customer satisfaction and profits, (2) supplier profits, and (3) company profits. How?

Walker advises that you must review the entire chain and use process mapping (or value stream mapping) techniques to predict the chain's performance in areas such as business system cycle times, inventories, distribution requirements, costs, risk/delays, and so on. His basic definition of a supply chain requires three parties as participants in supply chain improvement. Focus is given to the performance of the entire chain, not just any individual participating member. It's essential that we look at the performance of the system, where the system in this case is defined as the aggregate performance of three companies in the chain. Improved consensus in decisions is built through the members of the supply chain, internally first, then externally through the other members of the chain. Key to this consensus building is creating a Mandate Team within each company in the supply chain, which is made up of a champion in each of the following areas:

Design-ability

Produce-ability

Afford-ability

Sale-ability

Other appropriate capabilities

Each champion on the mandate team works to develop the internal consensus for a given issue, and then works with the other members in the supply chain to develop external consensus among the members of the chain. They must also have sufficient knowledge of the issues to be able to make decisions. A key element in the methodology is that *each member of the team* **must** be at least 70% satisfied with the team's decisions and direction. This is a subjective measure, which is literally voted upon by the team members. Otherwise, the decisions must be re-evaluated until this 70% satisfaction is achieved for *each* member. He advised that peer pressure usually mitigates or eliminates problems with getting this 70% satisfaction level, even with any "lone ranger" behavior among obstinate team members.

The mandate team consistently strives to identify the needs of the ultimate consumer of the products of this chain, and to meet or exceed those needs. According to Parker, this is paramount in supply chain management. Parker also observed that the benefits from supply chain integration are typically split one-third to the customer, one-third to the principle member of the chain, and one-third among the other members/suppliers of the chain.

Team members must committed to: (1) cooperation, communication and trust, (2) using the competencies of the supply chain, (3) generating long-term benefits, (4) common measures of success, (5) continuous improvement, and (6) sharing of competitive pressures within the extended enterprise. Surveys are often done among the team members to assure these commitments are being met, and if discrepancies are observed, corrective actions are taken to minimize them.

Walker reported benefits that included 20% to 70% quality improvement; 30% to 90% shorter cycle times; 15% to 30% waste reduction; three-fold or more technology gains; and diminished hazards through shared risk. Again, benefits were typically shared one-third to the end customer, one-third to the principle chain member, and one-third among the other chain members.

Walker also suggested that there are specific axioms (self-evident truths or propositions) for supply chain management. He indicates that these are patterned after the work of Boykin, Zampino, Doyle, Parker, and CAM-I-1995, and offered them as follows:

1. There is a shared specific focus on satisfying their common end consumer.
2. There is an alignment of vision.
3. There is a fundamental level of cooperation and performance to commitment (trust).
4. There is open and effective communication.
5. Decisions are made by maximizing the use of the competencies and knowledge with the supply chain.
6. All stakeholders are committed to generate long-term mutual benefits.
7. There is a common view of how success is measured.
8. All members are committed to continuous improvement and breakthrough advancements.
9. Whatever competitive pressures exist in the environment are allowed to exist within the extended enterprise.

He identified four companies who had agreed to participate in a supply chain improvement effort, and who after grading themselves on a scale of 1 to 10 on all 9 axioms found areas of substantial agreement and of substantial disagreement. These differences were then analyzed and used to improve their alignment to their strategies and goals.

A measurement template was presented for measuring supply chain performance, both internal to each company and external to the entire supply chain. These included financial measures such as revenues and assets employed, costs, inventories, and so on; quality, including customer satisfaction and defect rates, scrap and rework; cycle time, including order to delivery, product development time, internal raw material to finished goods, chain response rate, and so on; and technology, including enhancements to products, and product/process enhancements for each link in the chain.

These principles seem entirely consistent with Lean Manufacturing and in particular the Toyota Production System. Supply chain principles can be applied to dramatically improve business performance of the chain, and ultimately the satisfaction of the final customer.

Supply Chain Management: Another Model[3]

Martin Christopher takes a different, yet compatible, approach to supply chain management, as described in his book *Logistics and Supply Chain Management*.[3]

Christopher's key messages are:

- Focus on the customer, aligning products and service to them, working from them back into your company
- Fully understand all costs throughout the supply chain, not just manufacturing costs
- Centralize strategy and policies, but manage their deployment locally
- Constantly improve your understanding and performance

These key messages are entirely consistent with lean manufacturing principles.

Differentiating and Managing Customers

Regarding customers, he advises us to apply the Pareto rule (e.g., the 80/20 rule that 20% of your customers will provide 80% of your profits and/or volume, and/or some other critical measure), and provides an example of what you might find in your business as shown in Table 11-1.

Table 11-1 *Pareto Analysis of Products/Customers and Sales/Profits*

	%Products/Customers	% Sales/Profits
A List	20%	80%
B List	50%	15%
C List	30%	5%

It's fairly clear from the data in Table 11-1 that a business should likely focus on the A List for its products and customers. He also notes that it is routine to measure on-time, in-full delivery performance, or OTIF, but suggests adding to that the percent of product that was delivered error free. This could then be a measure similar to overall equipment effectiveness (OEE) within a plant and previously discussed, but as applied to customers, something that could be called Overall Service Effectiveness (OSE), where:

$$\text{OSE} = \text{on-time rate} \times \text{in-full rate} \times \text{error-free rate}$$

So, for example, if our deliveries were on time 95% on time, but we only supplied 95% of the requested volume on time, and we had a 1% return rate on that, then our OSE would be:

$$\text{OSE} = 95\% \times 95\% \times 99\% = 89.3\%$$

While each individual measure might seem reasonably good, cumulative performance looks more mediocre, and we would need to understand within our supply chain where the problems are so that we can improve our overall performance.

Christopher goes on to suggest that we manage our customers and products according to the models shown in Figures 11-1 through 11-3. Figure 11-1 takes the data from Table 11-1 and graphically presents a decision-making model for products and customers. The A List of products and customers represents those key accounts that you protect. While they only represent 20% of the total, they provide 80% of the profits/sales. The B List deserves further development, that is, the B products that serve A customers, and A products that serve B customers need more attention and potential investment and

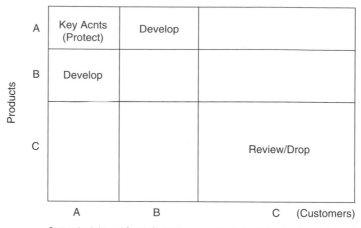

Figure 11-1 *Products/Customers Decision-Making Model.*

Figure 11-2 *Product Decision-Making Model.*

development. The B List wherein both customers and products are B's, deserve to be maintained. The C List deserves critical review and potentially culling from your business, especially the C customers buying C products.

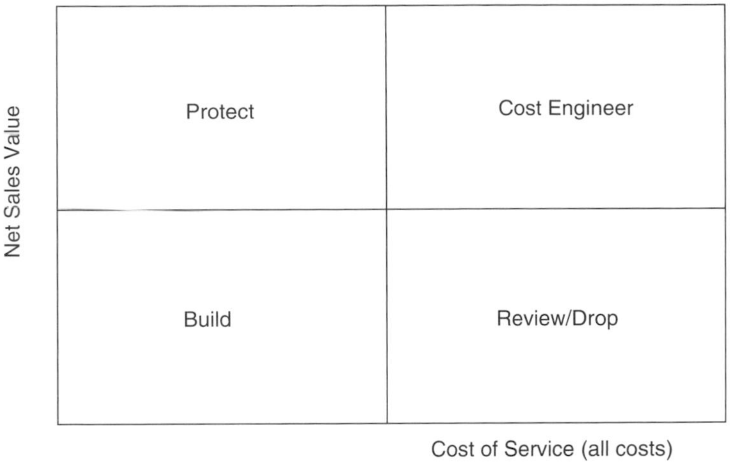

Protect	Cost Engineer
Build	Review/Drop

Net Sales Value (vertical axis label)

Cost of Service (all costs)

Source: *Logistics and Supply Chain Management* by Martin Christopher, by permission of Pearson Education, London, England, copyright © 1998. All rights reserved.

Figure 11-3   *Customer Profitability Model.*

Moore offers a similar methodology to this for aligning the marketing and manufacturing strategies, but goes into more detail on the methodology that could be used, including case studies.[4]

Christopher goes on to look at the business from other perspectives, as shown in Figures 11-2 and 11-3. Figure 11-2 provides a way of looking at volume and profit for each product, or Stock Keeping Unit (SKU), and then making decisions and developing a strategy for each product line. For example, if a product has low profit but high volume, the focus might be cost reduction; if it has high profit and high volume, then increasing its availability through improved manufacturing performance would be a focus (e.g., through improved OEE); if the product had high cost and low volume, then greater emphasis would be placed on just in-time delivery and minimizing inventory; if it had low profit and low volume, then the product might be discontinued, if there was no compelling or strategic reason to continue it.

He also uses a similar model to consider customer profitability using Figure 11-3. If the cost of service (including all costs through the supply chain) is low and the net sales value is high, protect these customers; if the cost of service and the net sales value are both high, then cost engineer for these customers; if the cost of service and the net sales value are low, then build and develop these customers. Of course, those that have high cost and low net sales value, should be reviewed and likely dropped.

Cost of Service

Clearly the cost of service would include the manufacturing costs, but Christopher suggests that we make sure other costs in the supply chain are also understood and managed, e.g.:

- Order processing
- Inventory carrying
- Packing and shipping
- Interplant freight and handling
- Outbound freight
- Commissions
- Pre-sales servicing
- Post-sales servicing

It's essential that we understand production and customer costs. The supply chain requires that we understand all costs: marketing and sales, production, delivery, and support, and map those costs through the supply chain, and then manage them. This reasoning is similar to the value stream mapping used in Lean Manufacturing.

Good Suppliers

Christopher characterizes good suppliers as:

- Willing to work in partnership or alliance
- Committed to continuous improvement
- Accepting of innovation and change
- Focused on throughput time reduction
- Committed to quality management procedures
- Supportive of benchmarking
- Flexible in helping design system logistics
- Sharing in common core values

- Actively seeking to improve communication
- Emphasizing primacy of quality/cost management

And he suggests evaluating your suppliers periodically, using these, as well as the following criteria:

- Pricing and value against the market
- Delivery performance: on-time, in-full, error-free; packaging; lead time vs. industry norm
- Cost savings initiatives
- Creativity: bringing in new ideas for improvements in supply
- Adherence to development schedules
- Response to quality problems; complaints; certificate to conformity/analysis
- Documentation
- Account, local, and technical support
- Administrative systems: invoicing, quoting, and so on

These criteria, and perhaps others that you consider important, should be routinely evaluated and used to improve supplier relationships and overall supply chain effectiveness.

Policy Deployment

Christopher, like many global company leaders, recommends using centralized global policies that are locally deployed. Global policies might cover the following:

- Network structure for production and transport optimization
- Information systems
- Inventory positioning
- Sourcing
- Trade-off analysis and supply chain cost control

Local deployment of these policies would allow greater freedom in the following areas:

- Customer service
- Market intelligence
- Warehouse management and local delivery
- Customer profitability analysis
- Liaison with local sales and marketing
- Human resource management

Functional Organizations vs. Supply Chain Organizations

Christopher offers a model for supply chain organizations and compares them to functional organizations.

Functional organizations tend to be:

- Silos, insulated from other groups
- Compete with other groups for resources, capital, executive attention, and so on
- Focused on inputs (costs, labor, and so on), not outputs: quality products and satisfied customers
- Examples of functional organizational behavior:
  - Sales seeks to close orders, with limited concern for production costs
  - Purchasing seeks to minimized purchased costs, not production costs
  - Production seeks to minimize product costs, not delivered costs

Supply chain organizations, on the other hand, tend to think at a higher, system level. Their characteristics include:

- Being horizontal: flat and de-layered
- Organized around processes along the supply chain, not tasks
- Built around multi-functional teams
- Focused on market driven performance measures (e.g., gross profit per product line, market share, and so on)
- Focused on:
  - Brand development, including new products
  - Customer development and management
  - Supplier development and alliances
  - Supply chain management: all elements of the chain

Supply chain organizations as depicted by Christopher are consistent with Lean Manufacturing principles.

Beta's Falcon Plant

Beta's Falcon plant was reasonably well run, and had made substantial improvement over the past few years. Costs were down, production was up, and overall performance was substantially better. In fact, its performance was a key factor in the turnaround of one division at Beta, and people were feeling good about their progress. One key measure, however, that they were faltering on was the equivalent of OSE previously described. At the corporate level, Falcon's performance was poor: 80% on-time, 95% in-full, with <1% returns, for a net OSE of 75%. On top of this, inventory turns were only average, running in the range of 5 to 10 turns. The plant manager felt it should have been well above 10.

This was disconcerting to the plant manager, and he investigated the reasons for these results. The investigation revealed that the plant's measure of delivery performance was 95%, and was based on product to the warehouse where the material was to be picked up by the customer at the loading dock. Purchase orders were freight on board (FOB) plant loading dock. But, several key customers were "saving money" by doing their own product pickups with their freight company, that was typically contracted by the customer.

Problems with their pickup included the freight company missing its route and delivery and the customer deciding to delay pickups because the product wasn't needed. Pickup delays by the customer were of little consequence to them, and in fact were a benefit; they were getting free storage and better cash flow at the expense of the Falcon plant. However, the Falcon plant was incurring reduced cash flow, higher storage and handling, and reduced return on working capital; the product was sitting in the warehouse, but payment was being delayed by the customer.

Beta, and the Falcon plant in particular, worked with key customers to get better overall performance in this area, that is, they created a more definitive set of requirements for pickup and delivery, including a cost for storage if pickup of the shipment was delayed beyond the purchase order requirement. This also forced the customers to stop using the Falcon plant as a buffer to their scheduling problems.

Beta's Tomahawk Division

Beta's Tomahawk division launched a major initiative in what it called supply chain management, the goal of which of course was to save money. Procurement led the initiative, consolidated suppliers, negotiated volume based price breaks, signed long-term contracts, and so on. At the end of the year, the Vice President of procurement reported having saved some $100 million, and presented several charts and tables indicating where the savings had been realized.

A review of Tomahawk's financials revealed a different perspective. Sales over the past three years had been relatively flat, costs had been relatively flat, and so profits had been relatively flat. Pricing had not experienced any dramatic shift, nor had market share. Where did the $100 million go? That amount of reduced costs should have appeared on the bottom line, given the business situation over the past three years. Apparently the money was saved with one hand, and spent with the other. For example, getting cheaper raw material that runs poorly and induces lower yields may *not* be the right thing to do. Getting new equipment that requires different spares and tools, and a new set of operating and maintenance knowledge may *not* result in any savings.

The lesson of course is that the Tomahawk division operates in "silos." It is a functional organization in Christopher's model, not a business system supply chain organization, and was using its so-called supply chain principles to squeeze suppliers on price. This approach often provides less than optimal results, and puts other functional areas of the business at risk. It often optimizes at suboptimal levels, impacting the entire business in a negative way. If you squeeze your supplier too hard, they'll have to make money some way, sometimes by eliminating things you need even more than the so-called price reduction. Focus should be given to business system performance using concepts such as supply chain performance, total cost of ownership, and net operating results.

Beta's Topmost-DPD Division

Of all Beta's business units, the Topmost-DPD division appears to be the one that does best in applying Christopher's policy deployment guidelines. Topmost has global policies covering:

- Marketing and advertising
- Production and distribution
- Common information systems
- Inventory management
- Sourcing
- Trade-off analysis and supply chain cost control

And, it also has local deployment of these policies, allowing for considerable freedom in the areas discussed in the following text. The Topmost-DPD division understands that in each geographic market, customers have varying languages, customs, cultures, laws, and over all expectations. Allowing for local deployment of its policies is a much better approach when accommodating all these local variations. Local deployment occurs in areas such as:

- Sales
- Customer service, including translation of product/service material into local languages and cultures

- Market intelligence regarding that particular markets needs
- Warehouse management and local delivery
- Customer profitability analysis
- Human resource management, including pay, benefits, and hiring/firing consistent with local laws, customs, and markets

The Topmost-DPD division has been very successful in applying this model, making substantial gains (>40%) in sales volume over a four-year period, and of course in market share and profits.

Beta's Melvin Plant

Beta's Melvin plant had been implementing Lean Manufacturing for some four years, working hard, although less than perfectly, in developing their long-term thinking, understanding their processes, and engaging their people in applying the supporting tools for Lean Manufacturing. For example, they had a very clean, well-organized plant (5S). As a result of this, as well as excellence in maintenance and preventive maintenance (PM) compliance, they had high equipment OEE (TPM principle). Indeed, they had achieved substantial improvement in one of their key productivity measures: labor hours per unit of product produced (LPU). However, the CFO asked "If we're getting so much more productive, why can't I see it on the bottom line?" Further review indicated that some 70% of their cost base was purchased materials, while only 7% was direct labor. In essence they had spent four years optimizing 7% of their cost base using principles and resources that should have been applied to the 70% of their cost. In fact, some of the LPU improvement was due to conversion of "make" items, which were getting better through using lean principles, into "buy" items, which was increasing their material costs. Unfortunately, there was no serious application of supply chain management or use of lean principles at their suppliers to improve the yields and cost of the materials that went to make their products. It seems they had excellent application of Lean Manufacturing to a process that only addressed a small fraction of their costs and, hence, were not seeing any major improvement in overall business performance. They would have been better served to apply the supply chain principles previously suggested by Walker, including getting their suppliers to use Lean Manufacturing.

Beta's Weeksbury Division

Beta's Weeksbury division developed an interesting approach to supply chain improvement. Rather than directly apply Walker's model, they actually modeled each step of the supply process, from raw material through to loading for delivery. The process models the variability of each step in the delivery process, applying the Monte Carlo simulation techniques. The key measure of success for this effort was return on capital employed (ROCE). The model simulated several years of operation by applying complex iterative statistical techniques, accounting for the randomness and variability of each step of the supply chain. For example, the model included the variability of the supplier, and their delivery system, as well as Weeksbury's variability in its ability to receive and use raw material, its ability to produce and warehouse, and its shipping system (the timely arrival and departure of transport). The model allowed the identification of those key steps in the process that had the greatest potential impact on improving ROCE. The results were surprising. It turned out not to be the manufacturing plant, but rather an area that most people would not have guessed (e.g., disruptions and variability in the raw material delivery system, in terms of quality, quantity, and timeliness). The Weeksbury division is now using the system to better understand where the major opportunities are in its supply/delivery chain. As it improves each, the inputs to the model will be changed for continuous improvement.

Summary

The models and examples discussed are consistent with Lean Manufacturing, but add detail to how you can more effectively apply supply chain management to improve your business. Analyze your supply chain from the customer/markets back to production and into your suppliers. Optimize the chain by analyzing your costs in your entire supply chain: suppliers, marketing and sales, production, delivery, and support costs. For each customer and product, know their gross profit contribution profile, which customers/products provide the most gross profit. Understand and routinely analyze which customers you want to protect, to develop, to maintain, and to eliminate. In applying this logic, develop the winners and cull the losers. As appropriate, create a mandate team in a supply chain: a key customer, you, and your key suppliers. As appropriate, use statistical models to model your supply chain's performance, and where the greatest leverage lies. Work within the chain to optimize the chain's performance. Develop strong supplier alliances based on the criteria previously suggested, and be very careful about squeezing your suppliers on price. Your suppliers have to make a profit; if you squeeze too hard, they may take something out that you need. Finally, measure and improve your customer service (e.g., on-time, in-full, error-free, or your OSE).

References

1. Parker, R., *Supply Chain Principles*, Newsletter, University of Dayton, Center for Competitive Change, May 19, 1999.

2. Parker, R., and Carlisle, J., *Beyond Negotiation-Redeeming Customer-Supplier Relationships*, New York: John Wiley & Sons, 1991.

3. Christopher, M., *Logistics and Supply Chain Management*, London: Financial Times/Prentice Hall Pearson Education, 1998.

4. Moore, R. *Making Common Sense Common Practice: Models for Manufacturing Excellence*, Boston, MA and London, England: Elsevier Butterworth-Heinemann, 2004.

Reliability-Centered Maintenance

A maintenance policy based exclusively on some maximum operating age would, no matter what the age limit, has little or no effect on the failure rate.

F. Stanley Nowlan and Howard F. Heap

Introduction

Reliability-Centered Maintenance (RCM) is a methodology for determining the most effective approach for maintenance. Effectiveness is determined by considering both reliability (or probability of failure) and overall cost. RCM focuses on those actions that will assure. (1) that equipment and systems achieve their inherent reliability and safety performance capability; (2) that proper standards are established for restoring equipment to functional capability when deterioration occurs; (3) that information is obtained for design improvements when inherent reliability proves inadequate for functional requirements; and (4) that accomplishes all this at a minimum life cycle cost.[1] With this focus, the primary objective of RCM is to *preserve system function.*[2] The methodology can be characterized as follows:

1. Identify the functions expected of your system (e.g., what's it supposed to do?) Make a list and be very specific, quantifying each functional requirement whenever possible.

2. Identify the failure modes can result in loss of system function (e.g., what's stopping it from doing that?) Make another list for *each* function.

3. Prioritize the functional needs using a criticality analysis of consequences and effects, for each function and each failure mode, (e.g., what's hurting you the most?) Make yet another list for each failure mode and for each function and prioritize based on "what hurts you the most?" in terms of cost, risk, production, safety, or environmental issues.

4. Select the applicable tasks, preventive maintenance (PM), or other actions that preserve or restore system function, that is, reduce the probability of failure or restore function, while being simultaneously cost effective (e.g., what are you going to do, on a priority basis, to assure or improve system reliability performance?). Note that some desirable tasks may be identified that would improve system performance. These are better characterized as redesign.

While simple conceptually, the analysis can be quite complex and time consuming. The complexity stems from actually following the steps for all the functional requirements, failure modes, frequencies, and criticalities, and then being disciplined about taking action to manage or avoid the failure modes. In doing the analysis (similar to a failure modes and effects analysis), *cross-functional teams are essential* to gather appropriate information on failure modes, frequencies, effects, and consequences, and to understand the system's operating context. In doing the analysis, process and equipment data/histories are essential for understanding these issues. In particular, the operating context can drastically alter the results of the analysis, even for identical equipment. A simple example of this would be two pumping systems that have identical equipment. One is required to operate in an emergency to support minimizing the risk of loss of life. The other is run continuously to support a production requirement in a process plant. Two identical sets of equipment have very different operating contexts and, thus, can have very different functional requirements, failure modes, frequencies, and consequences. The operating and maintenance strategy for each would likely be different, although some tasks might be similar. This is one of the reasons to be cautious about using vendor PM recommendations. While we should take their recommendations into consideration, we also must understand that vendors often don't understand the operating context of the system, which is essential in doing a RCM analysis and developing a proper strategy for system reliability, including related maintenance requirements.

RCM Standard SAE JA 1011

The above steps are a simpler approach to RCM. A more rigorous approach is offered by John Moubray,[3] and in Standard No. JA 1011 from the Society of Automotive Engineers, and is outlined as follows:

1. What are the functions of the asset in its present operating context?
2. How can the asset fail to fulfill each function?
3. What would cause each functional failure?
4. What happens when each failure occurs?
5. In what way does each failure matter?
6. What can be done to predict or prevent each failure?
7. What should be done if no suitable proactive task can be found?

The first four questions apply Failure Modes and Effects Analysis (FMEA); the last two identify the actions to be taken to manage potential failures. The fifth question determines how we should react to the failure. If the failure has little or no consequence relative to production, cost, safety, environmental hazards, or other risks, then run to failure may be acceptable, or even preferred. We'll also need to understand whether the failure is hidden or apparent to develop a proper response. These seven questions are best answered by a team of people who know the equipment best (e.g., operators and maintainers and perhaps designers and vendor representatives). The team is then led through the RCM analysis by a trained facilitator experienced in the RCM process and its application, and not necessarily the equipment expert, although that may be of substantial benefit to the analysis.

RCM analysis as traditionally practiced can be a tedious endeavor. Consider the aforementioned as a simplistic view of the steps in RCM. Before you can even do an analysis, you must specifically identify all system functional requirements and system boundaries. This of itself can involve lots of discussion and analysis. Then for each functional requirement you must analyze its failure modes, frequencies, and consequences, and determine its criticality; once that's done, you must determine the appropriate action to take to assure system function, in a cost-effective manner. Because of the tedious nature of the analysis—cross-functional teams spending days or even weeks on the analysis—it is sometimes referred to as a Resource Consuming Monster and Reliability-Centered Misery. Properly applied it need not be, and can be an essential element in manufacturing excellence. It has been successfully used in most industries, and has been particularly successful in aircraft and nuclear industries, both of which have a very low tolerance for "functional failure," because the consequence can clearly be catastrophic. And, aircraft companies can amortize the analysis cost over many of the same model aircraft, making it more cost effective to do the analyses required.

Expanding the process even further, the steps associated with doing the analysis as well as a logic tree for the analysis, are as follows[1]:

1. Identify the system and its boundaries:
    a. Inputs
    b. Outputs
    c. Resources
    d. Constraints
2. Identify subsystems and components
3. Examine functions, including the following characteristics:
    a. Primary or support
    b. Continuous or intermittent
    c. Active or passive
4. Define failure modes
    a. Potential failures
    b. Hidden failures
5. Identify effects and consequences of each of the failures

    a. Safety, health, environmental

    b. Production: availability, quantity, quality

    c. Capital cost

    d. Other

6. Analyze the criticality of each failure mode in its operating context

7. Develop a strategy and set of tasks that will assure functionality, detect or prevent degradation of function, assure restoration of function, or may require a design or operational change to achieve desired function

An exemplary RCM logic tree is shown in Figure 12-1.

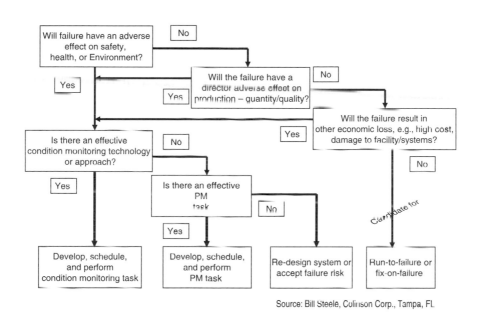

Source: Bill Steele, Colinson Corp., Tampa, FL

Figure 12-1 *RCM Logic Tree.*

An RCM Example

Let's *start* a simple example analysis to illustrate the potential complexity of doing an RCM analysis. Consider Figure 12-2.

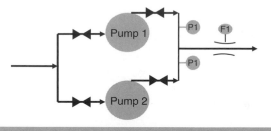

Simple Example: Pump 1 is primary; Pump 2 is backup
Functional requirement is 100 gpm @ 100 psi; each is capable of this.
1.   Would leaks be a functional issue? Leaking valves?
2.   Would calibrations or failed instruments be a functional issue?
3.   Would Pump 1 have the same failure modes as Pump 2?
4.   Do the two pumps have the same operating context?
5.   What if Pump 2 is in the storeroom?
RCM gets complex fairly quickly even for a simple system

Figure 12-2   *Example RCM Analysis.*

Here we have two pumps, either of which is capable of delivering our initial functional requirement, that is, 100 gpm @ 100 psi. Without actually following the steps in doing the analysis, let's just ask a few questions that are typical of those that might come up if we did a RCM analysis. This should illustrate its potential complexity.

We've predetermined that two of our functional requirements are 100 gpm @ 100 psi. How do we know the system is operating at 100 gpm and 100 psi? We'll probably need a flow meter and pressure gauges. Would 105 gpm be a functional failure? Most people say no, but it could, depending your system configuration, requirements, and boundaries. For example, suppose we had a small surge tank that overflowed quickly with high flow rates. Excess flow might be a functional failure, and we may want to include the surge tank in our system boundary. So, this series of questions might lead to a requirement to measure flow to a specific tolerance on the flow meter, of say ± 3 gpm. Suppose the flow meter has not been calibrated in three months, when its PM schedule says every month. Is that a functional failure? Most would say no, but it could be, depending on how we define a functional failure. If this were a pump in a critical process for a drug company, it would likely be a functional failure in that we have not met strict regulations for calibration of critical equipment, or it might not be. What about leaks: would a leak be a functional failure? It might. Suppose we were pumping a cyanide compound. A leak is most likely a functional failure. On the other hand, suppose we're pumping water, and using packing to seal the pump shaft. A small leak to lubricate the shaft might actually be a functional requirement, not a failure. We've only begun to ask a few relatively simple questions, each one of which will likely be debated, sometimes at length, and this scenario is for a very simple system. Clearly, an analysis could take a long time and become tedious, especially for complex systems. This in turn would require considerable patience.

Criticality Analysis

Figure 12-3 provides an example of a criticality analysis, this one for a process in an automotive plant. This is only an example, and you might have different criteria and standards for doing the analysis.

In this example, we must:

1. Specifically identify each function expected of the system.
2. For each function, we must understand the consequences of its failure.

Equipment Process	Function	Mode	Effect	Failure Consequence (frequency x severity)	Cause	Action Prevention or Detection

1. We must specifically identify each function expected of the system.
2. For each function, we must understand the consequences of its failure.
3. For each specific function, we must specifically identify its failure modes, and the probability of this occurring.
4. For each failure mode, its probability, and its consequence we must make a decision about the action that must be taken to manage the failure mode:
   a. Nothing—the probability or consequence is very low, or the cost of addressing it is higher than the consequence
   b. Changes—the design? Installation/startup practices? Operating practices? Maintenance practices? Training? HR policies? And so on.
5. We **must** follow through with procedures, standards, and so on, to act on the results.

Source: *Making Common Sense Common Practice,*
Elsevier Butterworth-Heinemann Books,
Boston, MA, & London, England

Figure 12-3 *RCM Criticality Analysis Example.*

3. For each specific function, we must specifically identify its failure modes, and the probability of this occurring (e.g., frequency).

4. For each failure mode, its probability/frequency, and its consequence, we must make a decision about the action that must be taken to manage the failure mode:

   a. Nothing—the probability or consequence is very low, or the cost of addressing it is higher than the consequence

   b. Changes—the design? Installation/startup practices? Operating practices? Maintenance practices? Training? HR policies? Other issues or policies?

5. We *must* follow through with procedures, standards, and so on, to act on the results.

We can use this model to assign relative values to things like frequency, consequence, detectability, and so on, to develop a scoring system for our criticality. We need not limit ourselves to the parameters identified here, and may want to add other issues like safety and environmental consequences. For example, a frequency of once per day might score a five, once per week a four, once a month a three, once a quarter a two, and once per year or less a one. For severity, one week of lost production might be a 10, one day of lost production a two, and one hour a 0.5. For detectability, a defect that is undetectable until failure might score a 10; difficult to detect might score a five, and easily detectable a one. In this model, you would multiply the scores in each category to get a criticality score, and then select a cutoff point for further consideration. Those cut offs might also be applied to each category, not just the total. As noted, you could also include other issues and score them as well (e.g., safety and environmental). Higher criticality scores would be given higher priority for action.

RCM and Developing a Maintenance Strategy

RCM data also strongly supports the concept of good design, installation, and startup; of routine condition monitoring; and of minimizing *invasive* time-based PM. For example, consider the data shown in Figures 12-4 and 12-5, the data is taken in part from the original

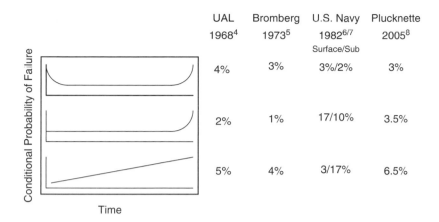

	UAL 1968[4]	Bromberg 1973[5]	U.S. Navy 1982[6/7] Surface/Sub	Plucknette 2005[8]
	4%	3%	3%/2%	3%
	2%	1%	17/10%	3.5%
	5%	4%	3/17%	6.5%

Figure 12-4   *RCM Age-Related Failure Curves.*

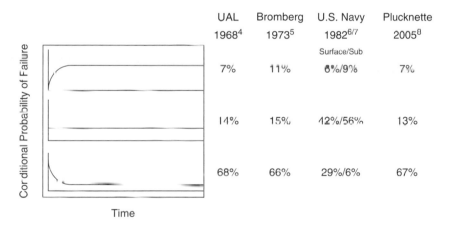

	UAL 1968[4]	Bromberg 1973[5]	U.S. Navy 1982[6/7] Surface/Sub	Plucknette 2005[8]
	7%	11%	6%/9%	7%
	14%	15%	12%/56%	13%
	68%	66%	29%/6%	67%

Figure 12-5   *RCM Random Failure Curves.*

RCM study by Nowlan and Heap,[4] and updated to include other and more recent studies.[5–8] What this data suggest is that unless you're operating a U.S. Navy surface ship or submarine, a time-based strategy for maintenance (e.g., time-based overhauls, replacement, and restoration), other than inspections and routine basic care PM, is not supported by the evidence for most applications. That is, the data indicate that only 8% to 13% of equipment (e.g., 4%+2%+5%) would benefit from a time-based approach to a maintenance strategy that limits equipment operating age, or requires replacement or restoration on a fixed-time or operating interval. The balance of 87% to 92% would not, but would substantially benefit from either excellence in design, installation, and startup to avoid infant mortality failures (66% to 68%), or from excellence in condition monitoring to detect onset of failure and then to manage the pending failure (constant conditional probability of failure). And note that if we do lots of time-based intrusive maintenance, we have a high risk of starting the infant mortality failures over again.

One point worth discussing is the substantial difference in the data for the U.S. Navy. They have much greater consistency in their practices and the rigorous application of standards. For example, if you are on a nuclear submarine, you **will** follow procedures. It's not an option. This rigor results in much greater consistency and fewer random failures being induced. It's a bit like the Toyota expectation for doing standard work. And they get better reliability. Finally, a point worth noting is that the 6% of failures being of the infant mortality type for nuclear submarines reportedly does *not* include those failures that occur during sea trials and commissioning efforts. It is only those that occur afterward when the boat has been released for sea duty. If failures during sea trials were included, the percentage would likely be substantially higher.

Nicholas reported the data shown in references 6 and 7 in a related paper published in 2004.[9]

Further, Aladon, LLC reported that for industrial equipment[10]:

- 5% require PM (time-based restoration and replacement)
- 33% require "detective" maintenance (inspection/test/check if it *has failed* [e.g., relief valves, protective devices])
- 25% require predictive maintenance (PdM) (check if it *is failing* [e.g., rotating machinery, transformers])
- 33% can be allowed to run to failure (no business consequence)
- 4% require re-design

So, according to Aladon's data, we should only have about 5% of our equipment on a time-based replacement or restoration maintenance plan. Most other equipment should be maintained according to its condition, which is determined by some form of inspection or condition monitoring, including operators observing process and equipment condition, not just PdM and intrusive inspections. About one-third of equipment could be allowed to run to failure; it has little business consequence. The notion that one-third of equipment in most plants could be allowed to run to failure with no business consequence is a surprise, and might be challenged by some.[1] However, the notion that one-third or more of common maintenance efforts might be modified or deleted is supported by the evidence, as discussed below.

Schultz, on analyzing some 20,000 maintenance tasks, found that PM tasks are often incorrect, as shown in Figure 12-6.[11]

As shown in Figure 12-6, an analysis of these 20,000 PM tasks for industrial equipment using the RCM methodology concluded the following:

Eliminate the task	29%
Use the task as-is	10%
Change task to PdM/condition-based	32%
Re-engineer task using simplified FMEA	26%

Moreover, *vendor* PM recommendations may not be accurate for your specific operation for a given set of equipment. As we discussed, the vendor may not understand your operating context, and can be understandably influenced by a desire to sell spare parts and services. Consider the work of Turner, summarized in Figure 12-7.[12]

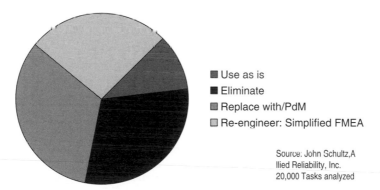

■ Use as is
■ Eliminate
■ Replace with/PdM
☐ Re-engineer: Simplified FMEA

Source: John Schultz, A
llied Reliability, Inc.
20,000 Tasks analyzed

Figure 12-6 *Analysis of PM Tasks Using RCM Principles.*

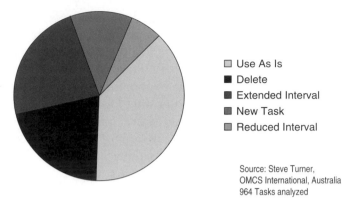

Source: Steve Turner,
OMCS International, Australia
964 Tasks analyzed

Figure 12-7   *Vendor PM Analysis.*

Referring to Figure 12-7, Turner used RCM principles to analyze some 964 vendor PM tasks, and recommended the following:

Eliminate	21%
Use as is	10%
Extend task interval	23%
Reduce task interval	6%
Add a new task	12%

This is not at all to suggest that vendors are providing poor information. Vendors are essential to our business success, work hard to support their customers, and are making recommendations with the best information they have. Unfortunately, we cannot and should not expect a vendor to understand our operating context and system requirements. They likely do not have the resources to provide that service for everyone in all operating contexts, unless of course we are willing to pay for that service. As such we should use vendor recommendations, but modify them as appropriate for our operational and business requirements, applying RCM to optimize their recommendations.

All of this suggests a rigorous analysis of maintenance requirements in the operating context of the equipment and is likely to be more optimal than continuing to do what you've always done. Think of the savings that might be available or the additional resources create if you did *not* have to do 20% to 30% of your current maintenance tasks. And, those that you did were optimized to improve equipment life, and/or to detect problems early allowing them to be better planned and scheduled to minimize their consequence and improve overall performance.

Potential RCM Pitfalls and Suggestions

Some potential RCM pitfalls are discussed below. A good practitioner will address these:

- RCM analysis typically has a slow start as people learn a new way of thinking; it can be very resource-intensive and time-consuming, so *patience is essential.*
- It's critical that you have cross-functional teams do the analysis; operating context and maintenance practices must be reviewed (e.g., running vs. idle, including startup and shutdown issues).
- Processes and discipline must be in place to *implement* the results of the analysis: operator checklists, TLC, work management/CMMS, PM procedures, planning and scheduling, trained personnel, measurements, and so on.
- The focus must be on defining effective maintenance and operating practices that will deliver the inherent reliability; once this is done and reliability is still inadequate, then we must change the design and/or further change our operating practices.
- Sometimes the focus tends to be primarily on PM activities. It's essential to use a more proactive, cross-functional team approach, which must include:
  - Design practices
  - Production practices
  - Procurement practices
  - Installation practices
  - Commissioning practices
  - Stores practices, and so on

- Looking at existing tasks and applying RCM to them as part of the overall approach.

- Using existing equipment histories and known failure modes for your initial efforts, before looking for all possible failure modes or those that you haven't experienced. In other words, don't do RCM on everything or go searching for all possible failure modes until you've addressed the known ones. Doing so is proper, for example, for NASA or Boeing, but may not be as useful for most applications or plants.

Some practitioners of RCM may strongly disagree with the last point, and so it deserves more discussion.

Most manufacturing plants do not have very good practices. For them to spend their resources analyzing all their potential failure modes, rather than focusing on their known failure modes that have a high consequence seems to be a non-optimal use of their very limited resources. The risk, of course, is that they may have a hidden failure mode with a low frequency of occurrence that could result in a substantial or even catastrophic failure. This is likely a risk worth taking for most plants. As their level of sophistication evolves and as they eliminate or better manage their known failure modes, shifting the emphasis to stress the proactive nature of RCM in looking for *potential,* low-frequency, high-consequence failure modes is probably more appropriate. However, an exception to this general approach would be if we're doing a hazardous operations analysis, or haz-ops analysis, and want to minimize the potential risk of an accident, then looking at *potential* failure modes is clearly the right approach. Minimizing the risk of safety, health, and environmental issues is also a critical design step that is routinely addressed by the design team, that is, they work to reduce the probability of occurrence of these events in the design phase. Further, if we consider NASA, the nuclear industry, or the aircraft industry, then doing an extensive analysis of all failure modes, both known as well as potential failure modes, is desirable. You can't get half way to the moon, be flying along at 35,000 feet, or find yourself in a meltdown, and say "Oops, I wish we had thought of this problem before now." In these circumstances looking at potential failure modes that have a very high consequence under various scenarios is essential, allowing for minimizing the risk of those failures. In most manufacturing plants, just addressing the known failure modes would put them way ahead of their current capability.

PM Optimization: RCM in Reverse

In an ideal world, we would use RCM to develop our operating and maintenance practices before and/or as part of the initial design and operation of the equipment. That generally does not happen, and often plants rely on historical practices and/or vendor recommendations. From the aforementioned data, these are likely to be ripe with opportunity. One method for capturing that opportunity is to do a PM optimization exercise on your known equipment PMs. It's similar to doing RCM in reverse, that is, rather than use RCM to develop our PM requirements, which is the best approach, we examine our PM requirements from an RCM perspective. Below are a series of steps that can be used to improve your PM practices.[10]

1. Select a system or equipment for PM optimization.
2. List this equipment's current PM practices and each PM's frequency.
3. For each PM task, ask, is the PM task necessary? That is,

   Does it help me detect onset of failure? (e.g., inspections, condition monitoring)

   Does it help to avoid onset of failure or extend equipment life? (e.g., oil and filter changes, lubrication, and so on)

   Is it related to avoiding a known, time-based and consistent failure mode? (e.g., the wear of the brushes on a DC motor resulting in a failure every 10 months)

   Does it help me find or avoid a failure with significant consequence?
4. If the task does not do any of these, could we eliminate it?
5. Does the PM have adequate specifications so that it's clear when we have a failure developing, or a non-conformance or corrective action that is required? For example, "Check belts" is not a good task description. A better description might be "check belts to assure they deflect no more than 1 inch under a 5 pound load at the midway point between the two sheaves;" or "check motor" likewise is not a good task description. A better one might be "check motor to assure that cross-phase inductive and resistive impedance imbalance is <5% at >90%

load." If the task does not have adequate specificity, then what specifications do we need to add?

6. What is basis for the current PM frequency? For example,

Equipment failure rate and history analysis?

Known failure rate?

Weibull analysis?

Best judgment?

Regulatory requirements, and so on?

For inspection and condition monitoring type tasks, what is your current "hit rate," that is, the percent of inspections resulting in detecting a problem requiring action?

Let's discuss this point. For example, if we do a PM inspection too often without finding a problem, then the risk that employees will not pay attention to the PM or even not do the PM at all is high. Some have suggested as a "rule of thumb" that an inspection-type PM should find a problem every 5 to 10 times the PM is done. Otherwise the person doing the work begins to think of it as a waste of time and does the task poorly. However, this ignores the consequence of failure. If the consequence of failure is very high, then people will be willing to pay more attention to getting it done properly. Further, it may not include adequate statistics for various types of equipment being reviewed, or the database may not be large enough to assume this rule of thumb to be adequate.

For example, suppose we're doing an instrument calibration once a week, and we've done this PM for 52 weeks now and only found one time when the instrument was not in calibration. Most people would say we're doing it too often and suggest doing it once a month or once a quarter. However, suppose further that the PM is for calibrating an instrument that measures the neutron flux that controls a nuclear reactor. Most people would suggest continuing to do the PM weekly and perhaps having a redundant instrument, and perhaps calibrating both even more often. Even this casual response may not be adequate. SAE Standard JA1012 provides statistical data for determining inspection frequencies with varying confidence limits. For example, if we want a 95% confidence of finding a fault in a timely manner, we would inspect or calibrate every five weeks. If we want 99% confidence, we would inspect every week. Considering risk and cost, we have to be able to identify the potential failure soon enough to act ahead of the failure. The cost of doing the task must be taken into consideration, both the actual cost of doing the work, and the potential consequential cost of not doing it, while balancing cost and risk.

Let's continue with our questions for PM optimization.

7. For each inspection, condition monitoring, or calibration PM, consider the consequence of failure to detect or avoid onset of failure. Note that a high consequence of failure should result in greater inspection frequency. Conversely, a low consequence of failure should result in a lower inspection frequency.

8. Based on your answers to questions 5 and 6 regarding failure rate, consequence, and so on, should any PM frequencies be increased or reduced?

9. Specifically look at any time-based intrusive PM (e.g., over-hauls, repair, and restoration). Could the need for this PM be validated (or eliminated or extended) by using condition monitoring to detect problems before doing the PM, or by using proactive methods to modify the design to improve equipment life? Define any actions needed and next steps.

10. Could certain basic care types of PM, such as tightening, lubricating, cleaning, inspecting, and so on, be done by operators to help avoid or detect onset of failure? Define the process for this. What issues or obstacles must be considered: safety, training, union, plant culture?

11. Could certain PMs be consolidated? That is, do a certain PM on one machine/process at the same time vs. re-visiting the machine/process frequently for "single" PMs.

12. Finally, are there other known failure modes where PM tasks need to be added to help detect or avoid these failures, particularly those which have a high negative consequence?

13. Re-define your PM requirements and, using this model, reassess them periodically and refine and improve this model and your PM process.

We can begin our application of RCM by examining existing tasks, providing better PM requirements, and making better use of resources in light of costs and risks. If we do this, we must understand that fundamentally we are trying to assure that our system functional requirements are being met in the most cost-effective manner, and that understanding and managing the failure modes that result in loss of function is the over-riding consideration. Also, as we will see later, having operators highly involved in equipment care is fundamental in this approach. It's also consistent with a foundational element of TPM to have operators involved in equipment care and maintenance.

Beta's Lowhansville Plant

Beta's Lowhansville plant decided to use RCM to improve its maintenance practices. Lowhansville had several people trained as facilitators in RCM, and for the first analysis had a consultant come in to facilitate it. The first analysis was attended by essentially all maintenance people, since it was viewed as a "maintenance thing." No operations people participated in the analysis. The equipment for the analysis was selected by maintenance and was based on downtime for that machine, and the operator's complaints about the machine not being reliable. The plant did not do a business-level FMEA to determine if this was the right machine to be analyzing from a business impact perspective, but the downtime data and the anecdotal complaints from operations made it an obvious choice.

The analysis resulted in a number of issues being raised, and several changes being made to the maintenance practices to reduce the risk of failure, and to better optimize the use of available resources. However, several failure modes were identified over which maintenance had little control. These were operator-induced failures related to startup, shutdown, and routine operation of the equipment. Several simple maintenance (or operator care) tasks were also identified that operations needed to do, above and beyond having more definitive and consistent operating procedures. Finally, a number of design issues arose that would help with improving the overall reliability of the machinery. Capital projects personnel were not present in the analysis either, so there was little support for these design changes.

Not withstanding the lack of operations and engineering support, the maintenance department put together a plan for proceeding with the changes identified and made some, albeit limited progress in getting these changes implemented; two of the on-site unions had a disagreement about some of the PM changes—who would do them, how, and when. The plant was highly reactive and so critical resources for re-writing the PM requirements were often not available, delaying the process for implementation. There was little budget for training people in the new requirements, and so even after they were put in place, they were often not done, in part because of a lack of skill and/or because of the reactive nature of the plant (people just weren't available to do the work). PM compliance was not good when the RCM effort began, and now because of some of the available resources were doing the analysis, PM compliance suffered even further, even for the new and improved methods.

Despite all of this, the maintenance department struggled on, believing that the methodology would eventually work—it was just a matter of time and patience in getting the process in place. Unfortunately, participation in operations and engineering in the analyses continued to be limited or nonexistent. Maintenance participation also dwindled as the process deteriorated. They struggled on for another six months after which the pressures of day-to-day operation overtook them and slowly over the next six months, RCM was gradually abandoned, ending up as a set of analyses in books on a shelf.

What went wrong? To sum it up, RCM was viewed as a maintenance thing, not about the plant and equipment functional reliability. Because of this, operations did not participate in any meaningful way, except to occasionally complain about people not being available to fix things because they were in RCM meetings. The follow up on the good things that were done was limited by inadequate procedural updates, inadequate training, and day-to-day reacting to the crisis of the moment.

Beta's Carr Creek Plant

Beta's Carr Creek plant had good support from its management for doing a series of RCM analyses. The plant was a process plant with several hazardous process streams and RCM was believed to be a means of improving maintenance and reliability practices, thereby reducing costs, while simultaneously reducing the risk of a major incident. The plant spent several months doing the analysis on critical systems using cross-functional teams with operations and maintenance people. Unfortunately, they spent so long doing the analysis the effort was so expensive that by the time it was done, there was no money left in the budgets to implement the recommendations, so they were never done. The simple lesson here is that if you're going to do the analysis, it had better be managed so that the results are implemented.

Beta's Oil Springs Plant

Beta's Oil Springs plant likewise decided to use RCM as a tool to improve its maintenance practices. For Oil Springs, several things were different from Lowhansville. First, they viewed RCM as a tool for supporting their TPM effort to help better define all the maintaining tasks that both the operators and maintenance technicians should do. Perhaps more importantly, maintenance was interpreted as maintaining the plant function, as per the TPM model, not as reducing the cost for fixing things, as per the Lowhansville model. Finally, the plant manager was fully supportive of RCM and using it to identify failure modes and consequences so that the business could be better operated. RCM was not a "maintenance thing."

Oil Springs likewise had a consultant facilitate the first analysis, and had several people trained as facilitators in RCM. The first analysis, however, was attended by a cross-functional team of operations, maintenance, and engineering personnel, since they recognized that many failure modes could be induced by any number of practices, including design practices. The equipment for the analysis was selected by operations, and they selected the machine that was having the greatest impact on production losses, developed from their OEE measurement. Oil Springs did *not* do a business-level FMEA either, because they felt they had sufficient data from the OEE measure and the market demand and gross profit data to make that determination.

The analysis resulted in a number of issues being raised and several changes being made to the maintenance practices to reduce the risk of failure and to better optimize the use of available resources. And, as with Lowhansville, several failure modes were identified over which maintenance had little control. These were likewise failure modes related to startup, shutdown, and routine operation of the equipment. Several simple tasks were also identified that operations needed to do, above and beyond having more definitive and consistent operating procedures. Since operating people participated in the analysis, they exerted a much greater degree of ownership and developed appropriate tasks for addressing these problems and avoiding and detecting the failures early enough to develop a plan for minimizing their consequence. Finally, as with Lowhansville, a number of design issues arose that would help with improving the overall reliability of the machinery. Capital projects engineering personnel were involved in the identification and resolution of these problems, so there was much greater ownership and support for these design changes. While they ultimately only did some of the recommended design changes because of capital budget limitations, they had much greater success than Lowhansville.

Below is a summary of the task responsibilities that were assigned from the analysis. Of the total number of tasks identified:

Operator tasks	66%
Maintenance tasks	32%
Other department tasks	2%

Design changes were also identified for about 10% of the failure modes that would help improve overall reliability of the machinery. Most of these were related to equipment redesign, but a few were related to process redesign. This characteristic pattern in the percent of tasks identified being the responsibility of the departments indicated continued as they used RCM to analyzed additional systems. At last count they had analyzed some 25 systems.

What went right? To summarize, RCM was viewed as a support tool for TPM and Lean Manufacturing. That is, the process was focused on maintaining the plant and equipment function in a cost-effective way. Because of this, operations participated heavily in a very meaningful way; they took on two-thirds of the tasks identified to improve reliability and performance. Most of these tasks related to getting the basics right all of the time. The follow up was excellent because the plant manager was supportive and continuously asking about the progress and the results of the program. Procedures were updated, training was routine, and the day-to-day operation demanded the implementation of the changes identified.

RCM and TPM: The Relationship

As previously mentioned, a question sometimes asked is whether RCM and TPM are compatible. They are, and RCM can be used to optimize the application of TPM principles.

As discussed in Chapter 9, TPM calls for restoring equipment to a like-new condition, *or better*. This is essential. According to RCM data, as much as 68% of equipment failures are the infant-mortality type, that is, in the first 30 days or so of startup. Given this, it's essential that we verify like-new condition using a stringent installation, startup, and commissioning procedure covering processes and equipment. It's also important to make sure we shut down the equipment properly. Lots of defects can be introduced during improper shutdown, only to appear during startup. It's also important to understand each equipment's or process's failure modes (defects) and effects in order to take steps in its design, operation, or maintenance to mitigate or eliminate those failure modes. RCM can help us understand and address those.

TPM calls for operator care and involvement in maintaining equipment, a must in a modern plant. Operators must have empathy for the machines, and maintain equipment function. However, as demonstrated at the Oil Springs plant, the operator often needs to be able to call upon more advanced methods and technologies to solve problems, including RCM, Root Cause Analysis, and PdM, all of which provide a method for understanding failure modes and their elimination or management.

TPM calls for improving maintenance efficiency and effectiveness. This is also a hallmark of RCM; many plants make extensive use of PM or so-called PMs. While inspections and minor PMs are appropriate, intrusive PMs for equipment more often than not make equipment less reliable. RCM studies indicate that only about 11% of equipment would benefit from an age-related intrusive PM strategy, and that 89% would benefit from a condition-based approach to maintenance. RCM helps determine the most common and consequential failure modes, which PM and/or condition monitoring method is most effective, and which equipment deserves attention in design and operations. PMs are optimized.

TPM calls for training people to improve their job skills. RCM helps identify the failure modes caused by a poorly trained staff. RCM is highly supportive of TPM, since training needs can be more effectively identified and performed.

TPM calls for equipment management and maintenance prevention. This is inherent in RCM principles by identifying failure modes and avoiding them, via design, installation, operation, or maintenance practices.

TPM calls for effective use of PM and PdM technology. RCM methods help identify failure modes and when and how to most effectively use PM and PdM to detect or avoid failure modes.

RCM and Six Sigma: Common and Uncommon Characteristics

RCM and Six Sigma have several elements in common. Both are rigorous, engineering-based methods. Both are data- and fact-driven. Both often require cross-functional teams to be fully effective. Both are resource-intensive. Both place a strong emphasis on continuous learning and improvement, on process management, and on appropriate process design/redesign. RCM focuses on preserving system functional requirements and identifying the failure modes that result in loss of function, applying the Failure Modes and Effects Analysis (FMEA). Six Sigma focuses on reducing variability, whatever its cause, and places more emphasis on customer feedback, analysis of variance, design of experiment, and the use of statistical process control methods for its analysis and improvement, applying the DMAIC model (Define, Measure, Analyze, Implement, and Control). Both are excellent tools, but RCM seems more appropriate for analyzing equipment-related problems, whereas Six Sigma seems more appropriate as a general analysis tool for reducing process variability.

Summary

Beta's experience is that you should get the basics right first; you don't need RCM to know that you need excellent operating practices: standards, checklists, startup and shutdown procedures, operator care, TLC, and so on; that you need precision maintenance (e.g., precision alignment and balancing in critical rotating machinery, work management, and excellence in your lubrication practices—clean, proper lubes, in the right quantity at the right frequency). At the same time, RCM is an excellent tool for helping optimize these practices through understanding the process and equipment's functional requirements, and then using RCM/FMEA techniques to better apply them.

Before doing RCM, make sure you have management commitment to RCM as a process for maintaining plant and equipment function, and to have the right people (operations, maintenance, and perhaps others) committed to do the analysis. Understand the time-consuming nature of the process and the need for patience. *Follow through* on the results—have the budget and systems to support the results of the analysis—don't let RCM be a book on a shelf.

Focus RCM on the most critical systems, generally using existing equipment histories and failure modes for your initial efforts; focus your efforts on the known and most consequential failure modes. Potential failure modes may be important to be addressed, especially for industries that have a high consequence of failure (e.g., hazardous process chemical plants, aircraft, nuclear power, and NASA), but for most plants just getting the basics in place is a good place to start.

You may not want to try to identify and prevent every possible failure mode, especially the ones that have low probability and low consequence. And some failure modes that have a high consequence, but very low probability may not be worthy of any operations or maintenance action.

RCM must be a "living" system that constantly seeks improvement and updates to failure modes, probabilities, and consequences, and incorporating other methods in problem-solving as appropriate. Properly applied, RCM can help most organization move to the next level in performance, and is a must in developing an overall long-term reliability strategy to enhance production and maintenance capability, safety, and cost effectiveness.

Beta will use RCM for optimizing its operations and maintenance practices. Where this is insufficient to provide adequate reliability, design changes to the process and equipment will be made.

References

1. Steele, W., Reliability Centered Maintenance Manual for University of Dayton Course Material, Dayton, OH: 2005.

2. Smith, A., *Reliability Centered Maintenance*, New York: McGraw Hill, 1992.

3. Moubray, J., *Reliability Centered Maintenance - RCM II*, New York: Industrial Press, 1997.

4. Nowlan, F.S., and Heap, H.F., *Reliability-Centered Maintenance*, United Airlines and Dolby Press, published by the Office of Assistant Secretary of Defense, 1978.

5. Pau, L.F., Editor, *Failure Diagnosis and Performance Monitoring*, European Airlines Maintenance Study Group, Marcel-Dekker, 1981 (reported in 1973).

6. American Management Systems, under contract to US Naval Sea Systems Command Surface Warship Directorate, Age Reliability Analysis Prototype Study, reported in 1993 (using 1980s data).

7. Allen, T., *US Navy Analysis of Submarine Maintenance Data and the Development of Age and Reliability Profiles*, Submarine Maintenance Engineering, Planning and Procurement (SUBMEPP), US Naval Sea Systems Command, Portsmouth, NH, 2001. (using 1990s data)

8. Plucknette, D., *Reliability Magazine*, Volume 10, Issue 3, January 2005.

9. Nicholas, J.R., Procedure Based Maintenance and Mastering the Maintenance Process, International Maintenance Conference, Bonita Springs, FL, December, 2004.

10. Moore, R., *Making Common Sense Common Practice: Models for Manufacturing Excellence*, Boston, MA and London, England: Elsevier Butterworth-Heinemann, 2004.

11. Schultz, J., Internal Data from Allied Reliability Inc., Tulsa, OK, circa 2000–2005.

12. Turner, S., Internal Data from OMCS International, Melbourne, Australia, circa 2003.

Predictive Maintenance/ Condition Monitoring

13

You can't predict anything with predictive maintenance.

Ron Moore

Introduction

Let's begin this chapter with the above admonition: you can't predict anything with predictive maintenance, or PdM as it is often called. Why then is it called predictive maintenance? The source of the name is likely lost in the fog of history, and the name is at least 20 years old. What you can do with PdM is understand the condition of your equipment, and based on certain standards or measures, you can detect an anomaly or the onset of failure, and anticipate that the equipment will eventually fail. Based on the severity of the condition and the consequence of failure, you can determine an appropriate priority and course of action to minimize the risk and consequence of the pending failure. You cannot predict *when* a machine will fail with any particular accuracy, but you can identify when the probability of failure has increased. A much better phrase for PdM is Condition Monitoring (CM), a phrase that will be substituted as appropriate in this chapter.

Cost Reduction Benefit of CM

The most common form of industrial CM, vibration analysis, has existed since the early 1970s, and there have been any number of studies that have provided compelling evidence of its business benefit. One of the more recent sets of data available that provides a compelling case for the use of CM technologies is provided below in Figures 13-1 and 13-2, which shows correlation coefficients of maintenance

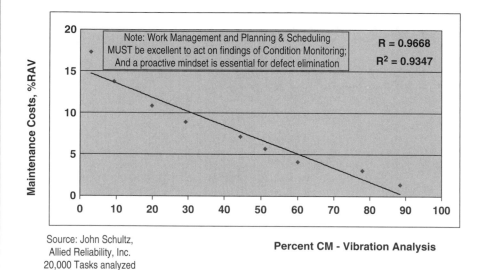

Source: John Schultz,
Allied Reliability, Inc.
20,000 Tasks analyzed

Percent CM - Vibration Analysis

Figure 13-1 *Correlation of Maintenance Costs vs. CM %: Vibration*

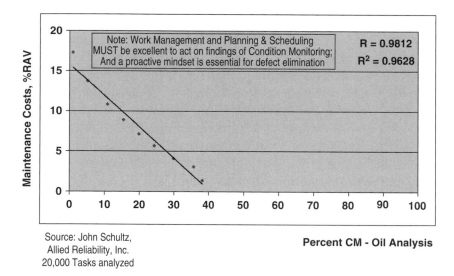

Source: John Schultz,
Allied Reliability, Inc.
20,000 Tasks analyzed

Percent CM - Oil Analysis

Figure 13-2 *Correlation of Maintenance Costs vs. CM %: Oil Analysis.*

costs and two common forms of CM (vibration analysis and oil analysis).[1]

Schultz reports these graphs as representative of these technologies for several industries (e.g., chemical, steel, pulp and paper, automotive, power generation, and general manufacturing).[1] And, he has similar data for other CM technologies (e.g., infrared thermography and other Non-Destructive Tests [NDT]). So Figures 13-1 and 13-2 should be viewed as characteristic for a wide range of technologies and industries.

As for the data itself, Schultz advises:

- The data is based on a minimum of 25 manufacturing plants in this industry and a minimum of five different companies in that same industry. The nine data points shown on the graphs are subsets that represent multiple plants that were divided into quartiles of performance and then plotted.

- Maintenance costs are normalized and presented as a percent of Replacement Asset Value (%RAV). The analysis included data for maintenance cost (including internal labor, contractors, materials, inventory, and so on); and RAV, which was validated by asset list and type.

- CM% refers to the percent of appropriate equipment on the CM program. Current CM coverage was reviewed for each asset and each technology. Then the asset list provided was put through a review process that generates a theoretical 100% coverage model, which is based on a failure mode mapping exercise. For example, if vibration analysis were applied to every machine that theoretically should be done from the failure mode mapping exercise, how many would that be? Then the number of machines actually being analyzed in the plant using vibration analysis was divided by the theoretical coverage model number. That results in the x-axis. It is not percent of critical equipment. For example, if the plant is in the chemical processing industry and if the requirement is to be in the lower end of first quartile coverage to generate the right amount of planned and scheduled work, then 60% of the 100% theoretical coverage is essential. Criticality dictates the 60% to pick. From this analysis for example:

  - CM%—Vibration is the percent of appropriate machinery actually on a vibration monitoring program.

- CM%—Oil analysis is the % of appropriate equipment actually on an oil analysis program.

- R is the correlation coefficient. While correlation does not necessarily represent a cause and effect relationship, it can be in inferred. R^2 represents the *maximum* amount of cause and effect relationship that can be inferred between the two parameters.

Preventive Maintenance (PM) and Increased Costs

Interestingly, PM, time- or cycle-based maintenance activities, result in a different outcome, as shown in Figure 13-3.[1] According to this data, beyond having more than about 20% of your equipment on time based maintenance actually increases maintenance costs. However, a word of caution on this—the PM% should be driven by equipment characteristics and its operating context, something that should be analyzed using Reliability-Centered Maintenance (RCM). This data, however, is generally consistent with what we learned in Chapter 12 on RCM, that is, most equipment fails in a random pattern, not a time-based pattern, making condition-based maintenance typically the most effective approach. Taking equipment down unnecessarily to perform maintenance that is not yet required is very

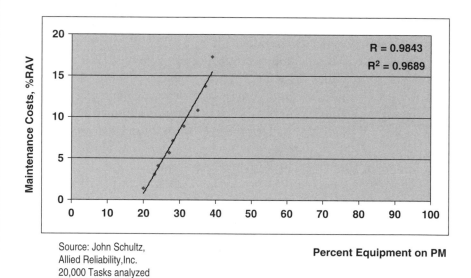

Source: John Schultz,
Allied Reliability,Inc.
20,000 Tasks analyzed

Percent Equipment on PM

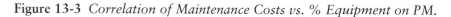

Figure 13-3 *Correlation of Maintenance Costs vs. % Equipment on PM.*

expensive, since production is lost and costs are incurred for unnecessary maintenance. "If it ain't broke, don't fix it." How do we know it "ain't broke"? We monitor it with the appropriate technology or method given its failure modes, at a frequency consistent with its failure frequency and consequence.

Common Mistakes in CM Programs

One note shown in the figures indicates that it is essential to have an excellent work management process, including excellent planning and scheduling in order to effectively manage the findings of the CM program. A common mistake made in CM programs is not having an adequate work management process, including a planning and scheduling system to follow through on the CM program's findings, so that corrective action is taken in a timely and cost-effective way. Additional detail on planning and scheduling is provided in Appendix A.

Another common mistake is not having all the appropriate machinery in the program. Very often plants will elect to only monitor those so-called "critical few" machines, leaving the others out of the program. Generally speaking this is a mistake. It's very difficult to live in two worlds: one where you care for some of your machinery, the other where you don't care for the rest, or you care much less. As shown in Figure 13-1 through 13-2, it's essential to have all of the appropriate machinery in the program if you want to operate in a lowest cost position. Further, getting the first few machines into the database is typically the most expensive part of getting started. After that, each additional machine added is likely to cost incrementally less

A third common mistake is not providing adequate training and skill development to the CM staff (e.g., technicians, engineers, and managers). Each person in the program should have a training and skill development plan, which is updated annually. Training, especially the initial training, can be intensive and expensive, but is essential for an effective program. As stated earlier, "If you think education is expensive, try ignorance." Technicians should be certified as practitioners in the skill required (e.g., Vibration Analyst Level I or II Tribologist Level I or II). While this certification will not guarantee competence, it will set expectations for learning and validation of that learning. The training and certification effort must be backed up by measures of the program's effectiveness. For example, measures might include the number of machines in the program, CM schedule compliance, CM action compliance (action taken before a failure occurred, and failure to detect) the number of times that machinery or equipment failed without detection. Costs for the program and overall costs should also be measured. If we're doing the right things, our costs should be declining, and our plant availability and reliability should be improving.

A fourth set of common mistakes is not having operators involved in (1) monitoring the equipment from a production process view, (2) helping select and schedule the monitoring of the equipment, (3) doing routine "look, listen, feel, and smell" for detecting problems, or not taking adequate care to avoid inducing defects. Operators cannot only detect defects, they can and must also take a very proactive approach to avoid inducing defects in the first place by improving operating standards and procedures. While it's common that CM technology can detect defects before an operator, having operations and maintenance work as a team to assure highly reliable equipment is essential.

Finally, a fifth common mistake is not having one person responsible for the conduct of the CM program and for the follow up of its findings. Distribution of the various technologies across a number of supervisors is *not* good practice and substantially dilutes the potential effectiveness of the program. Having one person responsible provides good alignment and communication regarding the structure and goals of the program and is essential.

CM: Understanding and Managing the Degradation Process

With CM we need to understand and manage the degradation process, so that we can avoid or at least minimize the risk and consequence of failure. Consider Figure 13-4.[2]

Condition, or resistance to failure, is on the vertical axis. Time is on the horizontal axis. Assume that we're monitoring the equipment using the appropriate technology or methodology for its failure modes, and at a frequency that is consistent with its failure rate, adjusted for the consequence of failure (e.g., a higher consequence might result in a higher monitoring frequency). For example, suppose this equipment is a pump, the failure of which would have a big consequence in cost and lost production to the business. Further, suppose that for this class of pumps we've been

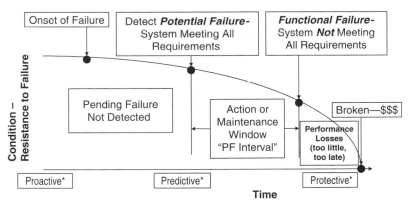

Sources: Ivara Corp, Hamilton, Ontario
*R. Baldridge, Cargill

Figure 13-4 *Understanding the Degradation Process. Avoid or minimize the consequence of failure.*

experiencing one to two failures per year. Given all of this, we would probably monitor these pumps once per month using vibration analysis, as well as having operators reviewing pump performance (i.e., cavitation, pressure drops, power levels, conformance to pump curves, and so on).

At the first point along the condition curve, we have onset of failure, but perhaps our technology is not capable of detecting this. It's still "hidden." At the second point, we detect a "potential failure," but our system is still meeting all of its functional requirements. This might be the random cavitation of a pump; we know a problem is developing, but the machine is still meeting its functional requirements for pressure and flow. At the third point on the curve is "functional failure;" the machine is no longer capable of meeting its functional requirements. At this point we may be operating at a reduced rate below the functional requirement or capability; we may be incurring quality problems or we may be inducing a potential safety or environmental hazard, but in any event, the system is operating at less than its functional requirements. Continuing to operate the machine will ultimately result in its total failure, and for critical equipment it's likely to have induced collateral damage by failing catastrophically and inducing additional costs and risks. If a "catastrophic" failure is of little consequence (little impact on production, costs, safety, or environmental compliance), then run-to-failure is acceptable and may even be preferred.

More typically, however, failures are substantially consequential. Referring to Figure 13-4 again, we want to use our knowledge of equipment condition to avoid or minimize the consequence of failure. The time between detecting a *"potential failure"* and a *"functional failure"* is called the P-F interval. This is the time during which we take action to avoid or minimize the consequence of the pending failure. These actions would focus on mitigating the pending failure and/or extending the P-F interval (e.g., tightening a fitting or lubricating a component). We might also make sure we have adequate spare parts on hand, and/or plan and schedule the repair so as to minimize its impact on production, and costs. Without this knowledge of equipment condition, we would more likely incur greater damage to the equipment, longer production downtime (and gross profit losses), and higher repair costs.

Notice also in Figure 13-4 there are three boxes along the horizontal time axis labeled proactive, predictive, and protective. Baldridge characterizes these zones as[3]:

1. Proactive: do those things that help you avoid or delay onset of failure. Examples of this include doing precision installation, startup and operation, and performing effective maintenance; doing TLC (tighten, lubricate, clean). In being proactive, it is particularly important for operators to operate precisely and care for the machinery to avoid failures. Having a design that is easy to operate and maintain helps. For example, we might apply poka yoke, or mistake proofing, in the design. Having precision maintenance is also essential (e.g., precision alignment, balancing, and fitups).

2. Predictive: monitor the condition of the equipment with the appropriate method or technology that allows you to detect a problem early enough to stop the defect, or to manage and minimize its consequence. While operators have a role here, maintenance and the use of CM technology tends to play a stronger role in this situation.

3. Protective: Suppose for some reason, you were unable to detect onset of failure and/or use the P-F interval to take action. Now the equipment has functionally failed. Now you are just trying to mitigate the pending failure so as to minimize its pending breakdown. Typically operators are more involved here.

PdM/CM Technologies

Let's examine the most common PdM or CM technologies, as well as other methods for CM. Again, these technologies allow you to monitor the condition of the equipment, including determining if there has been any deterioration in its condition, or its resistance to failure, so you can make a judgment about its condition and the risk and consequence of its failure. The technologies will *not* allow you to predict the exact timing of a particular failure. We'll consider the following CM technologies:

- Vibration analysis
- Lubricant/fluid analysis
- Infrared thermography analysis
- Electrical and motor current analysis
- Acoustic ultrasonic analysis
- Non-destructive tests (e.g., radiographic tests, dye penetrant tests, and so on)

It's also critical that operators routinely do process condition monitoring to detect any deterioration in the process and/or equipment, so it can be mitigated or managed. Operators typically number 2 to 10 times more than maintenance technicians in an operating plant. Omitting them from the CM program or strategy eliminates a huge resource and opportunity for better performance. Examples of operator process CM include:

- Process parameter monitoring, for example:
  Production rates
  Temperature
  Pressure
  Flow
  Pressure drop
  Dimensional parameters
  Product quality

Chemical parameters

Other parameters

- Operator CM: look, listen, feel, smell. Bissett and Idhammar independently indicate that some 40% or more of defects can be detected using the human senses.[4,5] While there is debate regarding the timeliness and precision of the information obtained this way, it is essential to involve operators in looking after the equipment with any "tools" they have. However, you should not totally rely on operators for your CM. CM technologies generally detect defects earlier than the human senses, and allow more time for minimizing the consequence of failure. CM can also be an essential part of the commissioning process to verify equipment condition at startup. Further, operators can be and must be proactive, that is, operating the systems and equipment in a disciplined, precise manner so as not to induce defects that result in the failures occurring.

A benchmark number is that about 50% of all maintenance work should be based on asset condition.[1,2] This is also consistent with the data from RCM studies which indicate most failures are random, typically making CM the most important maintenance strategy. A simple technique for determining the priority for the work resulting from CM is severity of condition times consequence of failure yields priority:

Priority = Severity of Condition × Consequence of Failure

Figure 13-5 provides a matrix of technologies, methods, and the equipment to which these can potentially be applied.[6]

Let's consider each of the more common CM technologies.

Vibration Analysis[1]

A good vibration analysis program:

- Covers all critical rotating machinery (stops, slows, or impacts production; or potential safety or environmental hazard; 80% of rotating machinery)

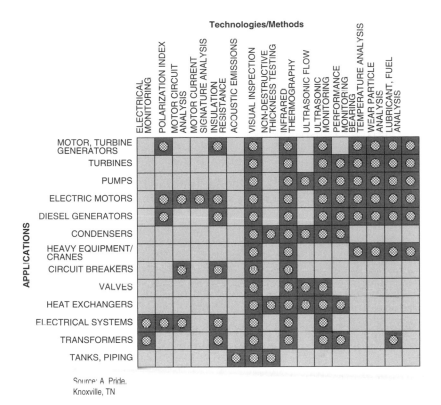

Source: A. Pride, Knoxville, TN

Figure 13-5 *Matrix of Equipment Types and Potential Technologies/ Methods.*

- Has current and adequate technology, both hardware and software
- Has adequate and skilled staffing that are certified as Level I, II, or III vibration analysts; a nominal estimate for the number of analysts required is a minimum of one analyst for every 400 to 500 machines.
- Uses RCM principles to optimize frequency and system applications
- Has the ability to analyze appropriate failure modes
- Data collection routes are streamlined to optimize analyst productivity
- Technology and skills are routinely updated and maintained
- Has a firm basis for assuring timely action is taken on pending failures—excellence in planning and scheduling and operations buy-in

Further, to properly collect the data and perform the analysis, you must also have good:

- Sensor selection
- Data collection techniques
- Hardware/instruments
- Software
- Analysis techniques (and training)
- Route selection for data collection and productivity
- Compatibility with other systems that will interface with the data collected (e.g., CMMS, networks, expert systems, DCS, and so on), as appropriate

Any number of vendors can help with these issues, and in most cases you will likely want to seek their help or that of other professionals, at least in getting started, and perhaps in actually running the program. Later in this chapter we'll review several things to consider regarding contracting your CM program.

The Basics of Vibration Analysis

Doing vibration analysis is similar to playing a musical instrument or even listening to music. For example, when you play an instrument (e.g., a piano), and you strike a middle C, it has a unique sound, because the string for this note vibrates at a specific frequency, its so-called natural frequency. The harder you strike the string, the louder (more energy) it will be, but it will still be vibrating at the same frequency. A different note would vibrate at a different frequency. Vibration analysis is very similar to this. You're "listening" (albeit seeing the "sound" on a graph) for specific frequencies and their amplitudes (energy levels) to determine the health of a given machine based on its physical characteristics. For critical machinery, it is essential to baseline machinery frequencies and amplitudes to determine machinery health and any deterioration over time. Two of the most common measures are:

- Time waveform: the actual motion of the machine in time
- Frequency spectrum: the motion of the machine at specific frequencies

Both can be very useful in determining machine health, and both have the time waveform as the source of the data for analysis, as illustrated in Figure 13-6.

Let's do an example of a vibration analysis using Figure 13-7. Let's suppose that this pump failed and we had to do a repair, but because of the urgency of getting it back on line, we had no time to balance the pump. If the pump impeller is not balanced, the result will be similar to an unbalanced tire on our car; it will "shimmy." This shimmy is induced by a heavy spot on one side of the tire, or impeller in this case, so that as the shaft turns, the heavy spot imparts energy through a tiny deflection of the machine. If we fixed a sensor on the pump casing to measure this energy, it would pick up a "pulse" each time

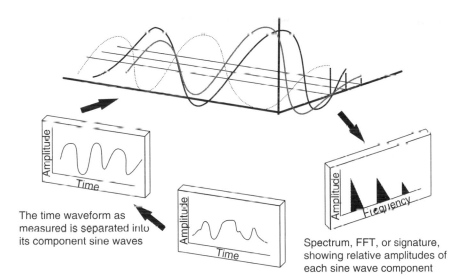

The time waveform as measured is separated into its component sine waves

Spectrum, FFT, or signature, showing relative amplitudes of each sine wave component

Source: Hewlett Packard App. Note No. 243-1, 1990.

Figure 13-6 *Relationship Between Time Waveform and Frequency Spectrum.*

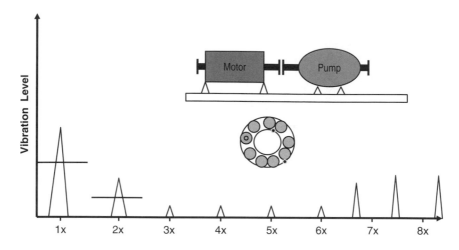

Figure 13-7 *Example Vibration Analysis for a Motor and Pump.*

the heavy spot on the impeller passed the sensor. The frequency of this pulse would of course be once per revolution of the shaft. This is shown as a "peak" in Figure 13-7 at "1x," or one times the shaft turning speed. The magnitude of the peak would be an indication of the magnitude of the imbalance.

Let's further suppose again because of time constraints imposed by production, we didn't align the coupling connecting the motor driving the pump impeller. In this instance, the shafts of the motor and pump would likely be offset or misaligned. This offset would be similar to having two heavy spots, one each on opposite sides of the shaft, and would induce a "pulse" of energy twice per shaft revolution, or two times the turning speed of the shaft, as shown in Figure 13-7. Again, the greater the misalignment of the shafts through the coupling, the greater the "pulse" of energy at 2x turning speed. It's hard to imagine not balancing new tires and aligning their car's front end when installing a new set of tires, and yet employees routinely fail to do this when installing or repairing the machinery that pays for their cars.

Let's further suppose we didn't have time to install the replacement bearings properly onto the shaft using the proper tools (e.g., induction heater, oil bath, press), and suppose we just hammered them on to the shaft and induced microscopic defects (the small dots shown on the bearings depicted in Figure 13-7). These defects would "ring" each time a single bearing rolling element impacted the defect, on the outer race (or circle in the picture) of the bearing at a lower frequency, and on the inner race at a higher frequency—same shaft speed but at a shorter circumference, and thus a higher frequency. The rolling element frequency might be somewhere between the inner and outer race frequency. The peaks at 3x, 4x, 5x, and 6x generally represent looseness, which shows up as vibration signals at multiples of the fundamental turning speed of the machine, or 1x. In this instance we're paying the power company extra for energy used to "beat our equipment to death."

To improve the reliability of the machine, that is reduce its vibration levels, we should set standards for each frequency shown and expect that machinery will be installed and commissioned to those standards, or better, and thus maximize the life of the machinery. One standard for setting these requirements is GM V1.0 2001.

In summary, with this technology we are "listening" for specific frequencies and their amplitudes, the result of which will tell us if our machinery is healthy or not. If not, we can take action to correct it, or minimize the pending consequence of its failure

While simple in concept, this practice is apparently difficult in execution, since most manufacturing plants, and Beta in particular, do not apply this technology as they should, to the detriment of their machinery's life, and to the increase of their costs, downtime, and production losses.

Crawford has written a comprehensive book on vibration analysis for those interested.[7]

Lubrication/Fluid Analysis

As we all know, if we don't properly lubricate our cars, they will wear out much sooner. If we don't plan to keep the car for a long time, then this is acceptable. Lubrication practices are simple and straightforward, but again, one of the most poorly performed practices in manufacturing plants. In a typical manufacturing plant, operator care and the discipline and consistency of operating practices and lubricating practices would quickly and inexpensively improve its performance. There are huge gains to be achieved in most plants by doing these two practices well.

Let's start with the basics: what is the purpose of lubricants? Lubricants[8]:

- Assure no metal-metal contact and subsequent wear or heating
- Inhibit corrosion/acids
- Inhibit oxidation
- Dissipate heat
- Inhibit sludge/lacquer buildup (in some instances)

Proper lubrication will assure that equipment lasts a long time.

For each machine, a specification is needed to provide its lubrication requirements:

- Viscosity and viscosity index (critical)
- Each additive and its required levels
- Maximum allowable wear particle levels: ISO grade cleanliness standards
- Any paraffins permissible (e.g., <1 ppm)
- Moisture permissible, if any
- Frequency of lubrication for each machine
- Quantity of lubrication for each machine

And then you must assure that it meets those specifications on a routine basis, including trending changes in the appropriate parameters and acting accordingly; and establishing alarm limits, lube change standards, equipment operating standards, and so on.

All lubricants should be consolidated, preferably to a single supplier that works with you to assure excellence in lubrication. Storage and handling practices must be superb so as to avoid contamination with dirt, moisture, or other contaminants.

Beyond having standards for each lubricant, both new and unused lubricants should be monitored or tested for contamination. For example, referring to Figure 13-8, Troyer and others report that some 25% of brand new lubricants straight from the vendor are ISO Grade 17 or worse in wear particle contamination.[9] Others report that this is as high as 100%.[10] This is caused by inadequate manufacturer quality control, and internal and external contamination (e.g., dirt ingress, condensation, various leaks, open storage containers, outside storage, and so on). Having clean lubricants (e.g., ISO grade 12 or better), is essential for long equipment life. Some of Beta's plants filter all of their new lubricants with a 3-micron filter to minimize wear particle contamination and machinery wear. Others appear unaware of the importance of lubrication excellence. If you want to assure clean lubricants, it is likely best to set up your own filtration and storage program appropriate to the requirements for your machinery.

It is essential to train your lube technicians in lube procedures and requirements. Each lubricating technician should be trained at a minimum as a Level I tribologist, and preferably a Level II. A key point regarding the greasing of bearings is *not* to over-lubricate, a more common problem than under-lubricating. Bearing vendors and tribologists can provide advice regarding this issue. Each technician must follow good sampling procedures and use proper disposal methods and handling for old lubricants. Sampling and monitoring must be done for all critical systems, e.g.:

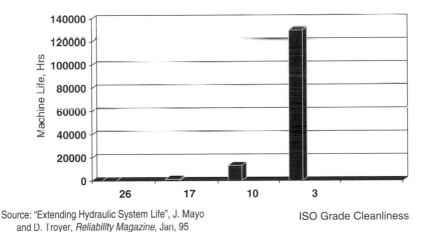

Source: "Extending Hydraulic System Life", J. Mayo
and D. Troyer, *Reliability Magazine*, Jan, 95

ISO Grade Cleanliness

Figure 13-8 *Effects of Wear Particle Concentration on Hydraulic System Life. Precision is critical to hydraulic system oil cleanliness.*

- Lube oil systems
- Hydraulic systems
- Gearboxes
- Transformers
- Bearings
- Diesel engines
- Chillers

Finally, lubricant/fluid monitoring must be integrated with the other CM technologies, assuring teamwork in issues related to all of the machinery.

Figure 13-9 provides a hierarchy outline for establishing a lubrication/oil analysis program.[11] It would be appropriate to have a comprehensive audit done of your lubricating practices to identify any opportunities for improvement.

Table 13-1 provides a listing of lubricant/fluid analysis methods.

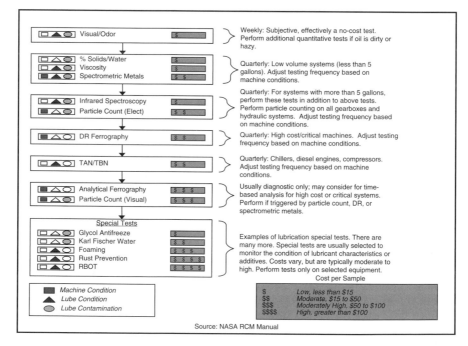

Figure 13-9 *Lubrication Analysis Hierarchy.*

Table 13-1 *Table of Lubricant/Fluid Analysis Methods*

Oil analysis methods to consider include:
Physical analysis
    Cleanliness, solids contamination
    Color, oxidation
    Atomic emission spectroscopy (< 10 microns)
    Approximately 20 different primary metals, indicating wear,
    contaminants, additives
    Ph: TAN/TBN
        TAN = tendency for acid corrosion
        TBN = resistance to oxidation
    Viscosity
    Fuel dilution
    Water contamination

<or>

Wear analysis
    Direct reading ferrography (relative: <or> 10 to 15 microns)
    5 microns: normal polishing wear
    >25 microns: severe wear
    Analytical ferrography: excellent particle definition
        ferrous
        non-ferrous
        polymeric, shape
        expensive: only used when other methods indicate problem
    Micropatch (coarse particle definition)
    Particle counting (~5 to 40 microns)
    Scanning electron microscope (SEM)

Infrared Thermography[1]

Infrared thermography is a technology that allows for the measurement of temperatures and/or temperature differences, and the creation of thermal images in a given set of equipment, so as to understand its condition and decide if any corrective action is needed. The actual measure is that of the infrared radiation emitted or reflected from a surface. Figure 13-10 provides a picture of the electromagnetic spectrum, from which the thermal spectrum is used for thermal imaging. Figures 13-11 and 13-12 provide several images (the lighter areas are the hotter areas) that suggest problems with the equipment from the image (i.e., a motor running hot [heat kills motors by breaking down the insulation],

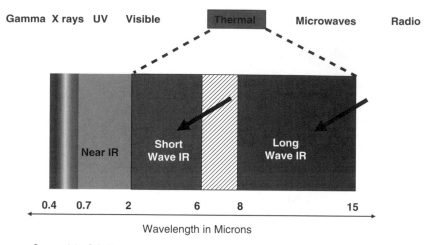

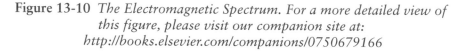

Wavelength in Microns

Source: John Schultz,
Allied Reliability, Inc.
20,000 Tasks analyzed

Figure 13-10 *The Electromagnetic Spectrum. For a more detailed view of this figure, please visit our companion site at: http://books.elsevier.com/companions/0750679166*

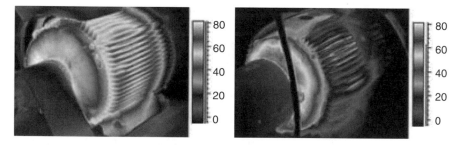

When comparing these two motors, the thermal patterns
are similar but there is a marked overall temperature rise
on the motor on the left.

Source: John Schultz,
Allied Reliability, Inc.

Figure 13-11 *Thermal Images of Two Motors. For a more detailed view of this figure, please visit our companion site at: http://books.elsevier.com/companions/0750679166*

Source: John Schultz,
Allied Reliability, Inc.
20,000 Tasks analyzed

Figure 13-12 *Thermal Images of Two Instruments. For a more detailed view of this figure, please visit our companion site at:*
http://books.elsevier.com/companions/0750679166

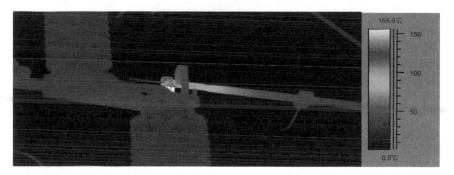

Figure 13-13 *Thermal Image of a Connector. For a more detailed view of this figure, please visit our companion site at:*
http://books.elsevier.com/companions/0750679166

a hot component that will have a shorter life, and a hot connector near an insulator on a power line). These are typical images from which a judgment can be made as to the equipment's condition, and the action necessary for avoiding failures, costs, and lost production.

The instruments used in thermal imaging and temperature measurement range from relatively inexpensive infrared thermometers for checking the temperature at specific spot locations and costing a few hundred of dollars to the much more sophisticated infrared thermal imaging cameras costing as much as tens of thousands of

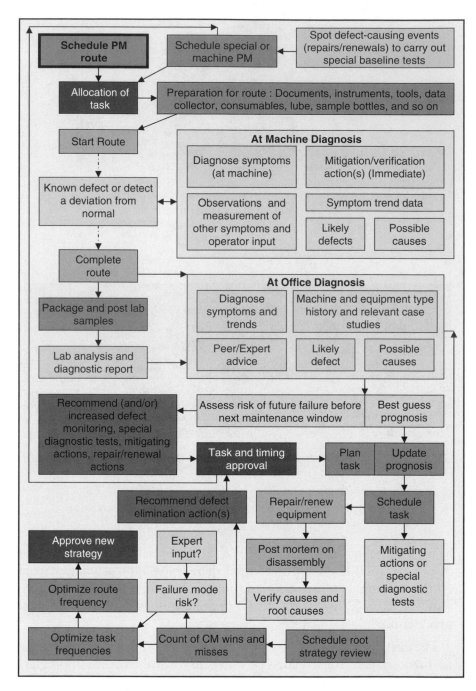

Figure 13-14 *Flow Diagram for CM Data Collection and Analysis.*[12]

dollars. They also have a similar range in their sophistication and capability. For example, radiometric devices have some emissivity correction and some have background correction capability, and are recommended for follow-up work or use with a non-radiometric imaging system. Infrared cameras provide the best information, but also require considerable training, particularly for the more complex surveys. It is recommended that surveyors be at a minimum certified Level I thermographers, and preferably a Level II.

A key parameter in doing infrared thermography surveys is to make sure the correct emissivity factors are used for the equipment being surveyed. Emissivity can be considered to be an "efficiency factor" for infrared emission. Values range from 0 to 1.0; a so-called black body emits all its infrared energy, and has an emissivity factor of 1.0; however, on the other hand, a perfect reflector has an emissivity of 0 and emits no energy. Real bodies are always in between, and change with material type, surface finish, wavelength, and temperature. Emittance changes with angle of view, and hence the angle of the camera or instrument relative to the surface being surveyed is important in getting good readings. It is critical that the emissivity factor be correct in doing the surveys. For example, a body with an emissivity of 0.2 at 400°F would emit the same amount of energy as a body with an emissivity of 0.9 at a temperature of 122°F, or 587 watts. Emissivity tables are used as guides, and measurements below 0.6 are unreliable, even when using emissivity correction factors. Common emissivities are as follows:

- Human Skin   0.98
- Water   0.98
- Electrical tape   0.95
- Paint   0.90
- Paper   0.90
- Copper (oxidized)   0.68
- Copper (polished)   0.02
- Aluminum (polished) 0.05

Convective cooling can also have a significant effect on the apparent temperature of a given component, and should be taken into consideration.

Safety can be a major concern when doing thermography surveys, particularly of electrical equipment, so appropriate safety rules must be applied. Suggestions for this are:

- Understand the risks and follow the guidelines of National Fire Protection Association (NFPA) 70E and Occupational Safety and Health Administration (OSHA) 1910
- Do not work alone and have qualified escorts with you as appropriate.
- Recognize that all voltages are potentially dangerous.
- Wear appropriate personal protective equipment.
- Do not break the plane of the door of electrical cabinets
- Minimize time spent in the so-called arc-flash zone

There are many applications for infrared thermography surveys to determine equipment condition. Table 13-2 provides is a listing of the more common ones.

Table 13-2 *Listing of Common Applications for Infrared Thermography Surveys*

Electrical
• Power Distribution and High Tension Lines
• Substation: transformers, transformer bushings, capacitor bank connections, disconnects, fuse clips, bus connections, relays and breakers, meter and control connections
• General Electrical Faults: breakers and contacts, bus and fuse connections, fuse clips, and stab connectors
• Motor and Generator Control Centers: connections and contacts, molded case air breakers, bus stabs and fuse clips, thermal overloads
• Enclosed Bus Runs: bolted connections, conductor fatigue
• Generating Stations: generator windings, generator brush riggings, generator feeders to primary, breakers and transformers, exciters, voltage regulators, relays and metering, boilers, associated steam piping, heat exchangers, radiators, cooling towers, motor control center (MCC), diesel engines and exhaust manifolds

Table 13-2 *Continued*

Mechanical

- Bearings and Couplings: alignment, heat dissipation, lubrication
- Energy Loss Due to Friction: conveyor belts, gear drives, belt drives
- Heat Causing Abnormal Stress: metals, plastics, glass, paper
- Non-Destructive Testing: gears, shafts, castings, extrusions, turbine blades, welds

Processes

- Remote Temperature Management
- Energy Conservation: heat loss survey, roof moisture survey, steam traps, heat exchanger inspection, building enveloping
- Quality Control: insulation void check, product inspection
- Refractory Studies: qualitative, quantitative thickness calculations
- Steam Systems Analysis: boiler studies (casing insulation, air pre-heater, feed water pumps, breaching and stack studies), steam distribution (main header insulation, steam turbine, traps, condensate system insulation, and overflow identification), process heating (exchanger fouling, condensate back-up, heat distribution, heat loss).

Motor and Electrical Testing[1]

There are a number of motor and electrical tests that can be done to determine the health of those assets. Common ones include:

- Insulation readings
  dielectric/polarization
  surge and high pot testing (recommended for acceptance testing only)
  megohmeter
  capacitance
- Conductor impedance
  resistance
  inductance

time domain reflectometer

motor current signature analysis

- Starting current and time
- Coast-down timing
- Circuit breaker inspection
- Power factor
- Harmonic distortion

Specific Motor Current Testing

Testing for broken rotor bars

- Applies most frequently to AC induction motors that are:

  heavily loaded

  incur frequent stops/starts
- Sampling: use 3,200 line spectra; Maximum Frequency of 80 Hertz (Hz)
- Tests must be conducted at >75% load
- Side-bands centered about 60 Hz, <60db difference from peak at 60 Hz indicates broken rotor bars

Vibration levels >0.05 inch/second at 120 Hz (60 cycle power) and 100 Hz (50 cycle power), generally indicates a non-uniformity in the motor magnetic field caused by:

- Eccentricity
- Impedance imbalance
- Rotor defect
- Stator defect

When buying new motors include a specification that limits maximum vibration levels in the motors to 0.1 inch/second at one times turning speed, and 0.05 inch/second at 120/100 Hz.

Motor insulation condition can be determined by readings/ratios with time:

- Megohm reading at 1 minute and megohm reading at 30 seconds
  Acceptable if: ratio is >1.5; Bad if ratio is <1.25
- Megohm reading at 10 minutes and megohm reading at 1 minute
  Danger if ratio of these is >5 (too dry/brittle); or if ratio is <2 (weak insulation)
- Acceptable if ratio is ~ 2 to 4

IEEE Standard 43 provides the latest information regarding motor insulation condition.

Airborne Ultrasonic Testing[1]

Airborne ultrasonic testing is essential for detecting leaks, including compressed gases, as well as vacuum systems. We can hear sounds with our ears between 20 to 16,000 Hz. Ultrasonic devices detect up to 100,000 Hz, and thus provide a wider range within which to detect leaks that would otherwise not be heard. With energy costs continuing to increase, finding and stopping leaks is a critical task. For example:

- A *single* 1/4 inch air leak (75 psi; $0.50 per cubic foot) costs over $1,500 per month, or *$18,000 per year*.
- A *single* 1/4 inch steam leak (160 psi; $10 per 1000 lbs. of steam) costs $3,000 per month, or *$36,000 per year*
- A *single* 1/4 inch water leak (60 psi; $1.50 per 1000 gallons) costs $1,200 per month, or *$14,400 per year*. And this doesn't include the waste water treatment and environmental costs that are compounded by the leak, and that are often higher than the water cost itself.

Leaks are very expensive. Find them, tag them, count them regularly, fix them again, and stop them in the future.

The types of instruments used to detect leaks include:

- Scanning devices (listening for high frequencies)
- Contact devices (touch the equipment), good for:

  Bearings

  Pumps

  Oil filled transformers

  Enclosed switchgear

  Waveform concentrator (dish)

Ultrasound advantages include:

  Directional

  Locatable

  Multiple applications

  Utilizable in all environments

  Early warning failure detection

  Supports other technologies

Techniques used to deal with competing ultrasounds include:

  Shielding

  Positioning

  Maximizing sensitivity

  Changing frequencies

  Creating barriers

  Closing doors

  Draping curtains

There are many applications for ultrasonic detection devices to determine equipment condition. Table 13-3 provides a listing of the more common ones.

Finally, ultrasonic detection can also be used to determine when a bearing needs lubrication. First, take a baseline measure of the bearing to determine its base decibel level. Following this, the following rules can be generally applied to determine lube needs:

- 8 dB over baseline indicates a lack of lubrication
- 10 to 12 dB over baseline indicates failure has begun

Table 13-3 *Listing of Common Applications for Ultrasonic Detection Devices*

Pressurized gas leaks (e.g., compressed air, oxygen, hydrogen, nitrogen, and so on)
Vacuum leaks
Heat exchanger leaks
Boiler leaks
Condenser leaks
Tank leaks
Pipe leaks
Valve leaks, seat, packing
Steam trap failures
Autoclave failures

Electrical equipment:
Arcing/tracking/corona
Switchgear
Transformers
Insulators
Potheads
Junction boxes
Circuit breakers

Mechanical inspection:
Bearings
Lack of lubrication and failure
Pumps
Motors
Gears/gear boxes
Fans
Compressors
Conveyers

- 16 dB over baseline indicates a failed bearing; replace it
- 35 to 50 dB over baseline indicates a catastrophic failure is pending; you do not need an instrument to tell you the bearing is failed.

Operator Process Condition Monitoring

As shown in Figure 13-4, it's essential that we involve operators in the care and monitoring of all of the equipment. One obvious reason for this is there are typically many more operators than maintenance technicians (e.g., 2 to 10 times more); the second reason is that they spend much more time directly with the equipment than maintenance technicians. They have intimate knowledge of the equipment's day-to-day performance. Operating practices when applied proactively will assure discipline and consistency of operation, and will reduce the number of defects introduced into the equipment, so that failures are avoided. At the same time, operators must also routinely monitor the equipment to detect any developing problems. As noted, operator process parameter monitoring might include:

- Production rates
- Temperature
- Pressure
- Flow
- Pressure drop
- Dimensional parameters
- Product quality
- Chemical parameters
- Other parameters appropriate

In applying operator monitoring we must define normal and abnormal measurement ranges and/or equipment condition, and then apply those to operator condition assessment. Trend charts and alarm limits can also be useful to detect deterioration in condition. It's also important to do the simple things—Look, Listen, Feel, Smell!

Finally, as with the CM technologies, it's essential that when a defect is detected that a process exists to: (1) correct it immediately, or (2) report the defect for corrective action; operations and maintenance work as a team to minimize and manage these defects. Essential to this is a work management process that converts the defect report into a work order that is planned and scheduled so as to minimize the consequence of the pending failure.

Finally, Figure 13-14 provides a model for collecting and analyzing your CM data.

Contracting CM

Before we consider specifically whether you should contract your CM program, let's consider some general guidelines on contracting. Contracting out a function or set of work is likely appropriate for[2]:

- High skill: low frequency work (thermography might be an example)
- Low skill: high frequency work (custodial work might be an example)
- Rapid startup or improvement in specialized (and sometimes poorly performing) functional areas (initiating a CM program, or improving a pump repair shop, might be examples)
- A "greenfield," prototype, or demonstration operation
- Emergency situations where the workload has overwhelmed existing capability
- For major capital projects or for overhauls/turnarounds to fill in short-term staffing shortfalls without incurring long-term employment obligations
- When needed to assure competitive position, e.g.:

  Work rule inflexibility with intransigent unions leading to a non-competitive business situation and risk of closing the business

  In a plant with several "bickering" unions regarding work rules and impacting productivity and the business success

- To help manage fluctuating work or product demand (and minimize potential impact on current employees)
- Any situation where business and competitive needs dictate a review of alternatives, especially for non-core work

You are strongly encouraged not to contract out your core competencies, those capabilities that if present substantially improve your potential for success, or if absent put the business at risk.

With this in mind, you might ask first whether any or all the CM program should or should not be contracted in light of the aforementioned criteria. It is recommended that you consider each separately first.

And then look at the program as a whole before making a decision, asking the following[1]:

- Are you big enough to support a good program?
- What skill level will be needed to provide the service?
- How often will you need the service?
- What's the risk of having it in-house, for example:

  Slow learning/benefit curve

  Future loss of trained staff to retirements/resignations

  Slower updates on the latest technology and skills

  It's not considered "core," and gets less respect, and/or is not sustained

- What's the benefit of having it in-house, for example:

  Higher degree of ownership and potential for sustained benefit

  Better communication with staff

  Lower perceived cost

- What's the risk of having it contracted, for example:

  Less ownership for the equipment and business

  Higher perceived costs

- What's the benefit of having it contracted, for example:

  Rapid startup and benefit achieved

  Constant upgrade to latest technology

  It's a core business to the contractor—your success is their success

Generally, the following is recommended:

- Large process plants: have a comprehensive in-house program with all appropriate technologies; you may still want to start certain technologies with a contractor, or have a contractor do a periodic audit of your practices
- Small plants (e.g., food, small chemical, and so on)—contract it
- In-between: case by case decision

- An exception: infrared thermography and any other high-skill, low frequency tasks—generally contract out except for the large operations (e.g., a large refinery or steel mill).

Beta's Boons Camp Plant

By just about any measure, Beta's Boons Camp plant was "dabbling" in CM. They had a vibration analysis program, but it was only being applied to about 10% of their equipment. They only had one full-time analyst and he had only one training course in the use of the technology, and was not certified to any standard for analytical skills (e.g., Level II Vibration Analyst). The planning and scheduling function *didn't*—it was more of a triage function for allocating resources to the most recent failure or crisis before it became worse; and there were times when the machinery was run to failure anyway. Their oil analysis program was similar (i.e., it was only applied to a few machines, technicians had little training, and there were no certified tribologists at any level). Schedule compliance for the lube PM and lube test sampling was better, but little action was taken when an oil sample indicated a problem. When the vibration analysis results indicated a problem, the equipment was often run to failure anyway, unless fortuitously there was a shutdown before that happened. The infrared thermography program was better, since an outside contractor certified as a Level II thermographer was used bi-annually for surveying the equipment and doing analysis; and because the failure modes were slower, there was in fact often time to avoid a catastrophic failure and/ or to minimize the consequence of the pending failure. An acoustic listening device was available (in a box in a storage closet), but was not being used for surveying for leak detection and correction. There just weren't any people for doing the surveys, and even if they did them, it was not likely that anyone would have acted on the results of the surveys; they only fixed the biggest, most obvious leaks, since most people were busy reacting to the crisis of the moment. Finally, the operations people did not do much CM. The general attitude about the equipment was "we break'em, you fix'em."

Boons Camp was making most of the common errors in a CM program, and the results, or lack thereof, were telling. Unplanned downtime was high, production performance was no better than mediocre, and millions in gross profits were being lost to frequent failures. However, with the arrival of a new plant manager, who was familiar with good CM, Boons Camp's management team was determined to correct this situation.

The initial action taken was to have the planning and scheduling function better staffed and trained, and to better manage the work they had. Planners were selected from existing staff, not new hires. The view was that better planning would improve productivity, and reduce the total number of technicians required. Planning was viewed as a foundational element in doing CM, that is, it was an enabler. Additional detail on effective planning and scheduling is provided in Appendix A. The next step was to add resources and people to the vibration monitoring effort, increase training, and monitor substantially more machinery. This took nearly one year to get in place. But the return on the investment was 10-fold—savings of 10 times the additional costs. The oil analysis process was enhanced, including the addition of an analyst and the expansion of the program to cover more machinery. A leak detection program was also put in place using an outside contractor, the same contractor that was doing the thermography surveys. The entire program was put under the supervision of a single engineer who was responsible for managing the program—the technology, the machinery being surveyed, the training, the defect reports, and in particular in coordinating with production to make sure they were working as a team, and that operators were involved in caring for and monitoring of the equipment. Improvements were substantial after 1 to 2 years, saving 10 times the additional cost, and freeing up other resources to do more proactive efforts. Additional improvements are being pursued.

Beta's Stanville Plant[1]

Beta's Stanville plant had a critical piece of equipment fail, and several maintenance personnel went "above and beyond the call of duty" and worked through the night to put it back into operation. The VP of Operations, sponsor of a reliability initiative, sent out an e-mail the next day thanking everyone for their extraordinary commitment to getting the equipment repaired and Stanville back on line.

Stanville's reliability engineer confronted the VP of Operations and asked if he was aware that the same mechanics that racked up a combined 172 hours of overtime fixing this critical machine had: (1) skipped the last three PdM and CM checks, and (2) was the worst performing department regarding PdM/CM compliance. The engineer also asked the VP when was the last time he sent out a thank you to a mechanic that got all of his PMs done on time, to the CM technician that completed a new certification effort, or to a mechanic that completed his precision skills curriculum? Unfortunately, it was a rhetorical question, since the VP had been rewarding reactive behavior, and essentially ignoring proactive behavior.

To the VP's credit, he became far more supportive of encouraging proactive behavior. He asked the engineer to organize a banquet for all personnel that met or exceeded the PdM/CM compliance goals that had been established. However, it's important that such an event not be viewed as a one-of occasion, or for that matter, not be done too frequently. For example, at Stanville, they decided to:

1. Recognize consistent proactive behavior, avoiding the rare stand-alone event.
2. Be consistent with the recognition, otherwise it can breed resentment.
3. Wherever possible, make it a group activity, with the criteria being as clear as possible.
4. Not make the reward too easy to get.

Whatever you reward or tolerate, you'll get more of it. In this case, Stanville was initially rewarding bad behavior—reactive maintenance; and had been tolerating bad behavior—not getting the PdM and CM work done. At Stanville the courage of one single engineer led to a dramatic change in the focus of the work, and the culture of the organization—rewarding good behavior in getting the PdM and CM work done, and being *in*tolerant of those who didn't.

Beta's Dwale Plant[1]

One of Beta's CM technicians was attending a conference where a CM expert stated that approximately 55% of all maintenance work should be done on a condition basis. That is, an analysis of the condition of the equipment and its ability to perform its function, as well as any anticipated deterioration of that ability, should determine when maintenance is to be done. Beta's technician thought this was "bull" and said so. The leader of the conference asked what percent of machines were on Dwale's CM program, and what percent of work orders were being generated from that, to which the technician replied 6% and 5%, respectively. The expert then asked if Dwale increased its percent of equipment to 50% to 60%, wouldn't it be reasonable that some 55% of work would flow from the CM program? The embarrassed technician meekly replied yes.

The Dwale plant was similar to the Boon's Camp plant—they were "dabbling" in CM, monitoring only those machines considered to be critical. However, there is much to be gained in reduced cost and waste in having a comprehensive monitoring program. The real benefit of a program occurs when you apply a comprehensive program to a very high percentage of your equipment. It's the only way by which you can detect problems early enough to plan and schedule the majority of your work, thereby maximizing equipment availability and minimizing production consequences, waste, and costs.

Beta's Watergap Plant

Beta's Watergap plant is quite a different story, with far better results. Watergap decided from the beginning to apply all the appropriate technologies to all the appropriate equipment. It provided an adequate staff with the appropriate training, all led by a single supervisor who was responsible for the results of the program. Annual refresher and upgrade training are routine. Planning, scheduling, and overall work management are excellent. Work is routinely coordinated with production and operators are an integral part of monitoring and caring for the equipment. Equipment downtime is <1%, while maintenance costs are in the benchmark range and overall performance is excellent.

Summary

From Chapter 12, we know that most equipment fails in a random pattern. Given the random nature of most failures, we must have a CM program that routinely detects onset of failure, so that we can make decisions about avoiding or managing those failures. Run to failure may be acceptable when the consequence of the failure is minimal, or when the cost of data collection is greater than the potential cost of the failure. For most equipment this is not the case, and so knowing the condition of the equipment puts us in a better position to make good business decisions.

CM, which is represented as a comprehensive CM program along with a good dose of operational support and process CM, is an essential tool for manufacturing excellence. As shown in Figures 13-1 through 13-2, the more comprehensive the CM program, the lower the maintenance costs. The material in this chapter can be used as a checklist for assuring a good program. A critical requirement for any program is having a good work management process for planning and scheduling the work that is identified through CM, as well as scheduling and complying with the CM and routine PdM tasks. Properly done, this will minimize costs and maximize plant availability and reliability. CM should also be included in the commissioning process to verify proper installation, as well as early detection and mitigation of problems to minimize the risk and consequence of failure. If you don't have the skill for a comprehensive program and/or think it will take too long to develop, it will likely be better to seek professional help by contracting the program, at least initially, and perhaps longer term.

Any good CM program should be based on the use of RCM discussed in the previous chapter. For a program to be effective and even optimal, it's essential that failure modes, frequencies, and consequences are known, so that appropriate CM tasks can be put in place to monitor those failure modes and then effectively manage them, including taking proactive steps to avoid them altogether.

Avoid making the more common mistakes in a CM program. First, monitor most of your equipment that has the potential to impact production, increase costs, or affect safety or environmental performance. Granted some will be more critical than others, but it's difficult to live in two worlds—one where you care and one where you do not. Second, establish a comprehensive work management program for planning and scheduling, both for the inspections using CM technology and the follow-through on the program's findings, so that corrective action is taken in a timely and cost-effective way. Third, provide adequate training and skill development initially to the technicians, and have ongoing refresher and improvement training, keeping your program at a state-of-the-art level. Each technician should be certified as a practitioner in the skill required (e.g., Vibration Analyst Level I or II and/or Tribologist Level I or II). Fourth, make sure operators are involved in monitoring the equipment and taking care to avoid inducing defects, working as a team with maintenance to assure highly reliable equipment. Fifth, have a periodic audit done of your CM program to help you identify opportunities for improvement and learning. Finally, have one person responsible for the conduct of the CM program, whether internal or contracted, and for the follow up of its findings. You will be much happier with the results.

Beta intends to follow these guidelines and apply CM in a much more comprehensive way, integrating it with routine maintenance planning and scheduling, and taking a much more proactive approach for defect elimination.

References

1. Schultz, J., Various Email and Verbal Communications, Tulsa, OK: Allied Reliability, Inc., 2006.

2. Moore, R., *Making Common Sense Common Practice: Models for Manufacturing Excellence*, Boston, MA and London, England: Elsevier Butterworth-Heinemann, 2004.

3. Baldridge, R., Email Communication, St. Michael, MN, Cargill Corporation, 2005.

4. Bissett, W., Verbal Communication, Newcastle, NSW, Australia, One Steel Corporation, Circa March 2004.

5. Idhammar, C., Conference Presentation, Raleigh, NC, Idcon Corp., Circa 1998.

6. Pride, A., Email Communication, Knoxville, TN, Pride Consulting, circa 1998.

7. Crawford, A., *Simplified Handbook of Vibration Analysis*, Knoxville, TN: Computational Systems, Inc., June, 1992.

8. Steele, W., Email Communications, Tampa, FL, Colinson Corporation, 2006.

9. Mayo, J., and Troyer, D., "Extending Hydraulic System Life," *Reliability Magazine*, January 1995.

10. Verbal information from various iron ore mining companies, Pilbara Region, Western Australia.

11. NASA Reliability Centered Maintenance Guide for Facilities and Collateral Equipment, February, 2000.

12. Todd, P., Email Communication, Melbourne, Australia, SIRF-Roundtables, 2006.

Root Cause Analysis

14

A pessimist sees the difficulty in every opportunity; an optimist sees the opportunity in every difficulty.

Winston Churchill

Introduction

Robert Nelms' has said that Root Cause Analysis (RCA) begins by examining catastrophic events, but ends up looking at the small things that cause the big ones. This lends credibility to the need to get the basics right first. It's the confluence of many little things done poorly that usually results in some big, bad thing happening. And, according to William Salot, RCA is any evidence-driven process that, at a minimum, uncovers underlying truths about past adverse events, thereby exposing opportunities for making lasting improvements. Let's explore this further.

There are many reasons why problems occur. For each reason, there are often many underlying reasons why the original event happened, and indeed in many cases a complex event tree can be created that describes at each level the additional underlying reasons for the previous event. As Nelms observed, catastrophic events more often than not result from the confluence of a number of small actions that ultimately led to the catastrophe. If any one of the actions that were a contributor to the confluence of circumstances had been eliminated earlier, the catastrophic incident would not have occurred. Fortunately this complexity minimizes the risk of major accidents or incidents. Unfortunately, it makes them more difficult to prevent, since any one event may seem innocuous or unimportant. RCA is a way of identifying root causes so that action can be taken to prevent them in the future. Figure 14-1 provides a simple logic chart for doing a RCA.[1]

The benefits of using a root cause approach are clear:

- It saves time by tackling the root cause first, not symptoms.
- A logical approach helps people discuss data and facts, not just opinions that are subject to much bias and change.
- It provides a means to collect and communicate facts and ideas.
- Most importantly, it facilitates finding root causes, so that action can be taken to avoid repeating the incidents that initiated the RCA.

The consequences of not doing RCA are also fairly clear:

- Organizations respond to symptoms, and so the problem isn't solved and reoccurs resulting in additional costs, lost production, injuries, or other hassles and frustrations.
- Organizations waste time implementing solutions for the wrong cause.

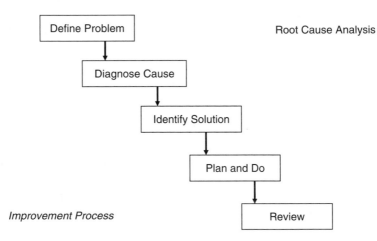

Figure 14-1   *RCA Logic Chart.*

Getting to the root cause of problems should be a way of life, something that is embedded into the culture of the organization. Unfortunately for most organizations, they tend to react to symptoms, that is, they're *symptomatic* in their behavior, instead of being *systematic* about getting to the root cause and eliminating it. A story that captures this behavior is that of the man who had a cough, so his friend gave him cough drops. After some temporary relief, the cough returned, and so his friend gave him even more cough drops. After some temporary relief, the man began coughing again, but more violently, eventually going to the hospital, where he was diagnosed with pneumonia, and was near death. After extensive hospitalization, treatment, and expense, and lost productivity, the pneumonia was cured. If only the initial treatment had not been so symptomatic and more systematic, at least some of this expense could have been avoided. And perhaps with a more proactive approach, it could have been avoided altogether.

Three Common Approaches

In this section we will review three different approaches to analyzing incidents to understand their root cause. Those three are:

- 5 Whys
- Cause and Effect Diagram with the Addition of Cards (CEDAC)
- RCA

There are other techniques, which are called by different names, and they may be as good or perhaps even better under certain circumstances, but those will not be covered here. The reader is encouraged to review and apply those techniques that are best suited for the problems at their organization.

5 Whys

This technique, which is simple to understand and straightforward in its application, is reported to have originated in the Toyota Production System. Their experience was that you should never accept the first reason given for an incident. Rather, you should always have a questioning attitude and probe until you fully understand the underlying causes of an incident. Their experience is that "Why?" needs to be asked at least five times before the true root cause can be established. Table 14-1 provides an example of 5 Why Analysis.[1]

While this one is fairly straightforward and does not affect overall policy, there are times when policy may be affected as a result of the analysis. For example, Table 14-2 provides another example that does in fact affect policy.[2]

In the second instance, substantive changes are required to the purchasing policies, including the method for evaluating their performance. While the analysis was straightforward, the policies and evaluation methods may require substantially more time.

Table 14-1 *Example of 5 Why Analysis*

1. Lost 300 units of production on September 3rd night shift—Why?
2. Conveyor stopped working—Why?
3. Drive bearings seized—Why?
4. Wrong lubricant was used—Why?
5. Unclear labeling resulted in the wrong lubricant—Why?
6. The manual had conflicting information for labeling, and technicians
   were not adequately trained in labeling. This is now actionable
   to eliminate the cause—modify the manual to resolve conflicting
   information; modify labeling to prevent future events; provide better
   training for technicians/supervisors.

Table 14-2 *Example of 5 Why Analysis[2]*

Situation	Countermeasure
Oil leak on shop floor Why?	Clean up oil
Machine is leaking Why?	Fix machine
Gasket deteriorated Why?	Replace gasket
Poor gasket material Why?	Change specifications
Purchasing saved money buying cheaper gaskets Why?	Change purchasing policies
Purchasing is rewarded for short-term savings	Change evaluation policy

In keeping with the KISS principle of Keep It Simple Stupid, using the 5 Why approach in routinely engaging the entire workforce seems to be a good approach. Try the simple way first, and if that is not sufficient or if it naturally leads to a branching of multiple responses to one of the questions as the 5 Whys are being asked, you may naturally move to a more sophisticated and complex approach. Before making the decision to use a more complex RCA model, a decision model for screening the more complex problems may be appropriate. Figure 14-2 provides an example of a screening tool for deciding when to apply more sophisticated tools for doing RCA.[3]

Let's assume that you have tried the 5 Whys and the results have not identified the root cause, or it naturally branched into multiple causes at one of the whys, or alternatively, the problem from the outset appeared too complex and it fit one or more of the criteria for your operation similar to those shown in Figure 14-2. As a result, you have decided to apply one of the more sophisticated tools. Two of these are discussed in the next sections.

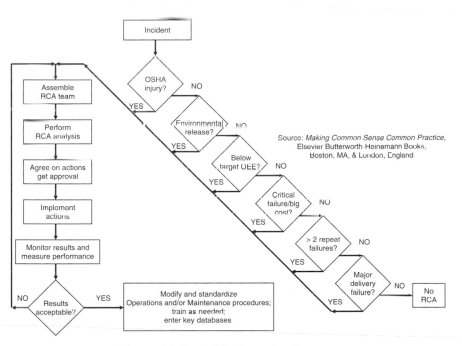

Figure 14-2 *RCA Screening Process.*

CEDAC[1]

CEDAC is based on the fishbone diagram originally developed by Ishikawa. It is typically best applied to solve chronic problems that have multiple potential solutions, versus being used to solve or get to the root cause of single events or major individual failures. Analyzing individual failures is best done using RCA techniques, discussed in the next section.

The CEDAC method includes a cause side and an effect side, which includes a problem statement and an objective or target statement. Potential causes and solutions are identified and overlaid, and the best solutions are selected. The CEDAC process itself uses the following steps:

1. Define and quantify the problem.
2. Determine your key performance measures.
3. Define and quantify a target statement.
4. Brainstorm all possible causes, and for example, use sticky notes to capture those.
5. Group potential causes onto the cause side of the diagram at each wishbone.
6. Brainstorm possible solutions for each of the identified causes.
7. Analyze and possibly test and implement each potential solution using a "dot" system to discern various solutions:

    Single Dot: The idea is of interest.

    Double Dot: Further investigation is required or to be tested.

    Triple Dot: The idea is being tested now.

    Or, after testing or further analysis:

    Double Dot: Draw a line through those double dots that are not viable.

    Triple Dot: Draw a box around those ideas that are successful.
8. Implement successful ideas.
9. Monitor performance.
10. Have systems—procedures, training, and so on, to maintain new performance.

Figure 14-3 provides an illustration of a simple CEDAC diagram, and will be used to address a problem involving poor fuel economy. The problem is defined and quantified (step 1) as poor fuel economy—the car is operating at 25 miles per gallon (mpg) vs. the manufacturer's report of 35 mpg. Our key performance measure (step 2) is mpg. Our target (step 3) is 31 mpg within the next four weeks; we suspect the manufacturer may have fudged the numbers a bit, or the circumstances for achieving the 35 mpg may not be realistic.

Continuing with this example, Figure 14-4 provides an example of the brainstorming exercise (step 4) where each possible cause for the poor fuel economy is listed.

Next, the possible causes are grouped into categories, as shown in Figure 14-5. In this case six categories were used: routes taken, car condition, car loading, fuel used, driving habits, and other miscellaneous causes.

Next, as shown in Figure 14-6, we brainstorm possible ideas for solutions for each of the causes (step 6).

In analyzing our potential solutions, we select, test, and possibly adopt these solutions using a "dot" system previously described to assess each potential solution (step 7).

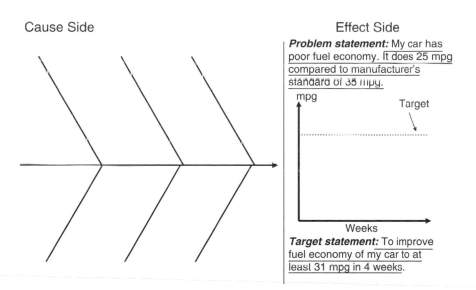

Figure 14-3 *CEDAC Example: Poor Fuel Economy.*

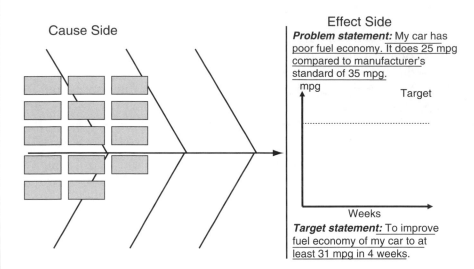

Figure 14-4   *CEDAC Example: Brainstorming Possible Causes.*

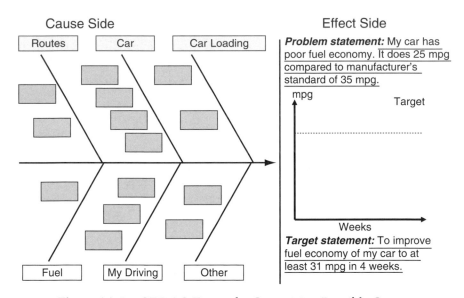

Figure 14-5   *CEDAC Example: Organizing Possible Causes.*

With the results of this analysis, we implement the successful ideas (step 8), continue to monitor performance for improvement (step 9), and put in place systems that will sustain performance (step 10). Figure 14-7 provides an example of this.

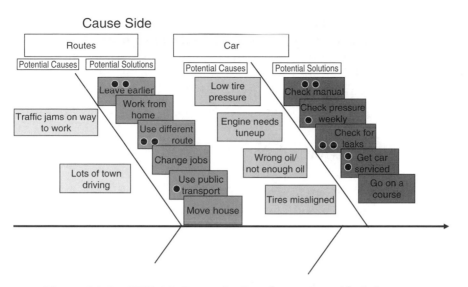

Figure 14-6   *CEDAC Example: Developing Possible Solutions.*

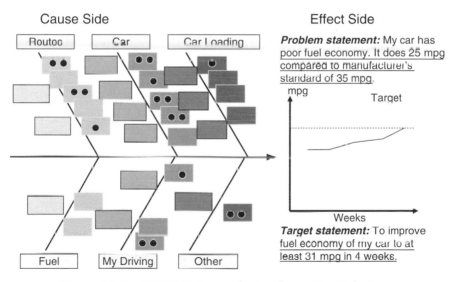

Figure 14-7   *CEDAC Example: Implementing Solutions.*

The CEDAC methodology is an excellent tool for brainstorming and solving chronic problems that have various possible causes. It requires that you bring together people with adequate knowledge of the problem to be able to contribute effectively to the brainstorming. It also requires that you be tenacious in developing, analyzing, and implementing the solutions. It further requires that you have systems to sustain the solutions implemented (procedures, checklists, and so on), along with measures that assure improvement and sustainability.

Apollo RCA Methodology[4]

The Apollo RCA methodology is similar to CEDAC in that it requires evidence-based cause and effect relationships to be analyzed. While it can be applied to all types of failures, it appears to work better for single event failures (e.g., an accident or a major individual equipment failure). It is reported to increase problem solving effectiveness substantially, and to result in substantial cost savings when applied. It can also serve as an effective platform for more creative solutions for the more complex problems. The steps for problem solving are also similar to the CEDAC methodology and are outlined as follows:

1. Define the problem:
   a. What?
   b. When?
   c. Where?
   d. Significance:
      Safety
      Environment
      Revenue
      Customer service
      Cost
      Other
      Frequency
2. Create cause and effect chart
3. Identify effective solutions
4. Implement best solutions

Figure 14-8 illustrates the cause and effect diagram and is the first step in applying the analysis. Using this model, we also look for *evidence to support each proposed cause*. If there is no evidence to support it, then it is must be sought out. We continue with the cause/effect analysis until we have reached the "point of ignorance," at which we can no longer think of any additional causes, we then begin to develop potential solutions, *from right to left*, challenging and analyzing causes and developing potential solutions to those, and working toward the primary effect. An example of this is shown in Figure 14-9.

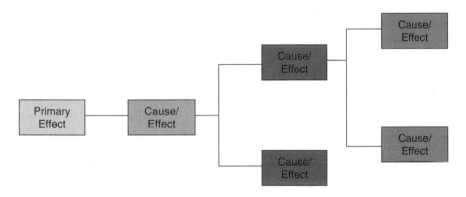

Source: Apollo RCA, ApolloniaPublications, Yakima, WA

Figure 14-8 *Cause and Effect Chart.*

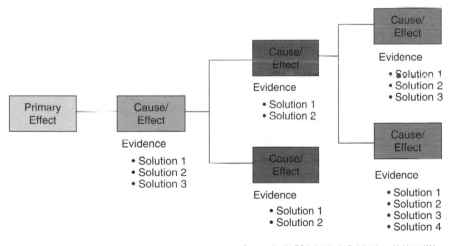

Source: Apollo RCA, Apollonia Publications, Yakima, WA

Figure 14-9 *Cause and Effect Diagram with Evidence and Potential Solutions.*

Gano observes that seeking a single root cause is a myth, and that rather what we seek is an understanding and ability to eliminate recurrence of the problem or event. For that there may be any number of solutions, any one of which will work, but of course some better than others. Solutions must:

Prevent recurrence

Be within your control

Meet your objectives, which should include: (1) not creating additional problems, (2) preventing similar occurrences, and (3) being cost effective.

The best solutions balance issues such as cost, risk, probability of recurrence, and difficulty. There is typically not one best solution, just as there is not one cause for most problems. We must eliminate the condition or action that combined to result in the event being investigated. He suggests avoiding solutions that say "review, or analyze, or investigate." This may be necessary, but does not represent true solutions, but rather a process for finding a solution. He also suggests avoiding use of the prefix "re" in solutions (e.g., retrain, review, replace, redesign). Again, this may be necessary, but is not viewed as a solution that addresses the cause. Finally, his data indicate that about 5% of problems have no solution, perhaps because of a lack of understanding of the circumstances that led to the problem. The good news is that 95% do have a solution.

RCA Example

Let's consider an example to illustrate the aforementioned methodology, in this case an accident that could have resulted in a serious injury or even a fatality.

First, the problem is defined as follows:

- What: potential electrical contact
- When: October 28, 1992
- Where: top of elevator, West Buchanan Building

- Significance:

  Safety: Potentially serious injury or even fatality

  Environmental: na

  Revenue: na

  Cost: total $13,650

  Frequency: 1st time

  Investigation: 100 hours

The cause and effect diagram is shown in Figure 14-10.

Results

In this case, the cause and effect chart demonstrates that the potential electrical contact (and serious injury or death) resulted when the electrician's hands were near the power as part of an annual maintenance activity. This was normal. Unfortunately, the circuits were energized. This was not normal. The reason for this was that the circuit lockout was incomplete because procedures had not been reviewed and/or upgraded to require this, *and* because a clerk had used the elevator earlier, with the permission of the electrician, to deliver the mail. This

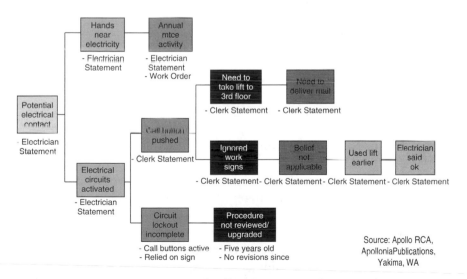

Figure 14-10 *Cause and Effect Diagram Taken to Solution.*

need occurred again, and the clerk chose to ignore the signs not to use the elevator because the electrician had earlier said he could use the elevator. A combination of events (previous use of the elevator having had approval, ignoring work signs, and an inadequate lockout procedure) resulted in an event that was expensive and could have been fatal.

And so it is with RCA, that often the confluence of several events or circumstances leads to the incident, any one of which were it not present, the incident would not have happened. The most cost-effective solution in this case is to upgrade the lockout/tagout procedure, and train the electricians to routinely apply the procedures and checks before proceeding with the work.

RCA Rt

Holmes offers a similar approach to the Apollo methodology previously described, but makes a direct link to 5 Whys as part of a continuum for RCA, and uses visual queues and software systems as an integral part of the analysis. The methodology is called RCA Rt.[5] The Rt is an abbreviation of Roundtable, whose members at SIRF Roundtables in Australia recognize the critical importance of defect elimination in the journey to high reliability at low cost. As a result, considerable discussion and research was put into problem solving and RCA. The RCA Rt process brings together the outcomes of this research and is presented below under sub-headings that deal with: (1) the development of a cohesive approach, (2) the importance of problem definition and follow through, (3) cause trees, (4) action management, and (5) incidents and ongoing concerns.

The Development of a Cohesive Approach

Most companies fail to develop a culture where problem solving is routine at all levels. One of the stumbling blocks to this is that companies often employ different problem solving and RCA approaches at different levels of the organization, in different parts of the organization, and for different problems. Very often the shop floor is encouraged to use one technique while engineers and managers are encouraged to use others. Different thinking is applied in different departments and for problems addressing personnel or equipment issues. While some differentiation is appropriate to the nature of the problem, some employees, including those who often have invaluable information, can become confused and/ or disengaged by the lack of consistency. This is a barrier to the development of a true culture of reliability and manufacturing excellence. A RCA screening process provides an effective way of insuring that priority and resources are placed on the appropriate problems. However, it can create a divide between the thinking applied on a day-to-day basis and the thinking applied on important or critical problems.

The RCA Rt methodology reportedly provides an interlocking process to handle this potential problem by accommodating various levels of complexity and importance. The process engages shop floor personnel to routinely think in terms of a structured problem-solving process, which includes 5 Whys on day-to-day issues. If the problem is more difficult or if it is critical, then more comprehensive tools are applied in a graduated way. However, the tools are reportedly designed to be a logical extension of the same process so the front-line people stay engaged throughout. Smaller organizations are able to handle this methodology with simple paper-based documentation. More complicated problems may require the use of software systems to assist where management of a large number of issues becomes important (e.g., storing information, documents, photos, managing actions, and reporting on progress). Such a system supports RCA in a similar way to that of Computerized Maintenance Management Systems support planning and scheduling.

A key ingredient in the process is the use of visual workplace concepts to guide people in determining how much effort should be placed on problem solving. This progression is illustrated as follows.

1. Front-line personnel are encouraged to apply 5 Whys on a daily basis and routinely ask "why" equipment failed or production targets were not met, rather than just putting the equipment back on line and forgetting about it.

2. If front-line personnel and their leadership feel that the problem needs further review, it is recorded. For example, a rule for this might be that if there is a production loss of more than 10 minutes, then the issue must be recorded. The model shown in Figure 14-2, could also be used for this decision making. They self-assess using a risk and consequence chart to define the consequences of the failure and its frequency. The system is designed to analyze the problem and then associate a risk level

with a particular color. The color coding indicates the minimum level of intensity that must be applied to the problem solving. Green might indicate low probability and low consequence, which would suggest a simple 5 Why analysis is the only requirement. Orange might suggest additional thinking on problem definition is required and the 5 Whys must be examined further to see if there are other possible "Whys," which leads intuitively to the development of a cause tree. A requirement at this point is that evidence must be presented against the identified causes in the cause tree to encourage personnel to more thoroughly investigate the issues. Other colors require an even more comprehensive analysis, including requirements that management be made aware of the problem, that target dates are set, that a cross-functional team is formed, and that evidence and actions are recorded and managed. Figure 14-11 provides an example of the color-coded risk matrix for determining the minimum level of effort that must be applied to a problem.

One advantage to this approach is that it is self managing so the process expands to suit the need in a logical and transparent way. Personnel themselves are identifying when more help is required and stay "on board" throughout.

Loss Type		Consequence				
		1 Insignificant	5 Minor	5 Moderate	5 Major	5 Catastrophic
Safety Risk-Injury		No Risk to Personnel	Medical Treatment	Possible Lost Time	Likely Lost Time	Potential Fatality
Production Loss Risk		~0	≥10 minutes	≥3 hours	≥week	≥1 month
Maintenance Costs Risk		~$0	≥$500	≥$10K	≥$100K	≥$1M
Likelihood	Likelihood Examples	Risk Rating				
A—Certain	Several times per year					
B—Likely	Annually					
C—Possible	Every few years					
D—Unlikely	Once in the life of the plant					
D—Rare	Not known to happen at this site					

Minimum Requirement for RCA
No Five Whys or RCA Required; no need to record concern
Single Person Five Whys
Team Based RCA
Complete Team Based RCA with Emphasis on Evidence
Complete Team Based RCA with Executive Participation

Source: Bill Holmes, SIRF Roundtables, Ltd. Melbourne, VIC, Australia

Figure 14-11   *Example Color-Coded Risk Analysis Matrix*

The Importance of Problem Definition and Follow Through

Holmes' experience has been that the most common reason individuals fail to solve problems is because they fail to properly define the problem. It is natural for people to remember a past situation and immediately try a solution found successful in the past. People naturally assume causes and apply treatments to symptoms rather than to the root causes.

Another observation is that very often investigators find that some people have known of the problem, understand the causes, and have identified the correct solution, but that the problem has not been treated for a variety of reasons. This situation can be due to personnel not being able to gain appropriate authority when cooperation between departments is required, something that is often accompanied by cynicism or frustration.

To minimize the risk of these difficulties, the RCA Rt process encourages personnel to think in terms of three phases, those being Focus, Find Causes, and Fixed Forever. These three phases are then divided further into steps that lead personnel through a simple but comprehensive process. The logic for this is shown in Figure 14-12.

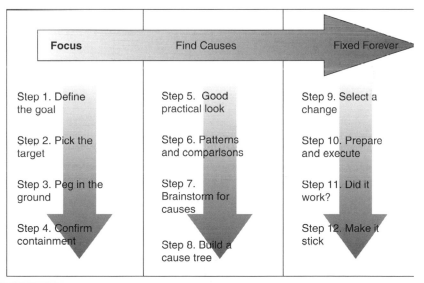

Source: Bill Holmes, SIRF Roundtables, Ltd., Melbourne, VIC, Austra

Figure 14-12   *RCA Rt Process.*

These steps can be further divided into useful tactics that help personnel work through the steps. This is shown in Figure 14-13. The graphics shown are often posted to help publicize and guide the thinking process.

Cause Trees

The Ishikawa diagram shown earlier in the section on CEDAC tends to gather ideas together in themes along each branch of the diagram. The cause tree on the other hand is more of a visual representation of the 5 Whys process. A branch on the tree illustrates a cause, continuing with identifying the cause(s) of that cause, and so on. The 5 Whys approach becomes a branch of the cause tree.

With this approach, people are able to come to grips with 5 Whys and use it routinely. Step 8 of the RCA Rt process encourages people to ask whether there might be another branch of "Why" for a given observation. So Fig 14-12 develops into Figure14-13. Asking further "Whys" leads to development of the cause tree.

Focus	Find Causes	Fixed Forever
Step 1. Define the goal.	**Step 5. Good practical look.**	**Step 9. Select a change.**
•The is, the should, and the gap	•Find out what happened.	•Remove,
•One-of event or on-going condition	•Look at the process.	•Neutralize,
	•Ask the people.	•New technology, or
Step 2. Pick the target.	•Ask what has changed.	•Eliminate the need.
•One plant,	**Step 6. Patterns and comparisons.**	**Step 10. Prepare and execute.**
•One process,		•Get authority.
•One problem.	•Visual tools	•Explain, execute, and fine tune.
Step 3. Peg in the ground.	•Yes/no table	
•Assess the impact.	**Step 7. Brainstorm causes.**	**Step 11. Did it work?**
•Keep a finger on the pulse, e.g., run chart	•Every idea is a good idea.	•Collect data and evaluate.
Step 4. Confirm containment.	•Check the checklist.	**Step 12. Make it stick.**
	Step 8. Build a cause tree.	•Let the system know.
•Does it stop the problem affecting people, plant, environment, or customer?	•Build the tree and check the flow.	•Install warning signals.
		•Learn from the solution.
•Look for side effects.	•Test for 'switch on and off' and actually present.	•Thank everyone who helped.

Source: Bill Holmes, SIRF Roundtables, Ltd., Melbourne, VIC, Australia

Figure 14-13   *RCA Rt Process.*

Experience has shown that this type of introduction to cause trees is very easy to understand and ensures that all personnel actively participate. Employees with more experience in cause tree development and particularly engineers who might participate in more complex or higher priority investigations may choose from a number of additional techniques to test the integrity of the cause tree. For example they may "turn off" a cause to see whether the outcome would be changed if the cause did not exist.

Experience has also shown that it is more important in the initial stage of cause tree development to keep the process simple and to avoid overly complex rules and syntax between causes in order to keep everyone engaged. After the tree is well developed and ideas have emerged then you can sweep back through and more rigorously test the logic in the tree.

One of the great strengths of cause trees is that it provides a visual representation of the different factors that lead to the undesired outcome. It is not essential to identify the most extreme limit of the branch or to treat the most remote cause. It may be that an intervention may be carried out on an intermediate cause that eliminates the possibility of the root cause leading to repetition of the problem. Step 9, Select a Change, uses this insight to encourage investigators to think of and choose from four different solutions. Most people initially think of new technology to overcome a problem. However, in many cases there are more cost-effective solutions that flow from the analysis of removing the cause, neutralizing its effect, or eliminating the need altogether.

Action Management

Success in RCA depends on evidence. It is not productive to try to solve solutions from the office. The various potential causes that have been identified must be investigated to ensure the assumptions are supported by hard evidence. Very often the actions that are needed to gather evidence are relatively small. For example, "Gather the broken pieces for the postmortem," "Save the post-trip data," or "Ask Joe about the incident." The actions are small, often short duration, but are essential. Establishing the root cause of a mechanical failure without a postmortem review can be a frustrating exercise.

The critical job of gathering evidence starts with gathering physical evidence *at the time the incident occurs* or the issue develops. For example, a failed bearing must be retained for inspection. Interviews with personnel are best done while memories are fresh. Simple statistical process control tools applied by personnel close to the problem yield excellent results.

It is not uncommon for organizations to see the importance of defect elimination, to encourage RCA, and to see early success only to see failure after several months. One cause of this failure is that leadership is distracted by other tasks and fails to support and encourage personnel in the actions required to find evidence and successfully conclude investigations. Success can mean several teams of personnel working on different root cause analyses. There may be several actions in each, so the job of supporting and encouraging can become difficult. It is desirable to document these actions to encourage and support progress. However, a Computerized Maintenance Management System can be a poor management tool for these small, high turnover activities. Paper systems can work, but software solutions may be an appropriate alternative to support cause tree development and action management.

Incidents and Ongoing Concerns

Personnel regularly think of RCA in terms of investigations into incidents or events. This response is perhaps due to safety management systems that encourage investigations into events like a vehicle accident or a fire. In fact, RCA is entirely appropriate for reviewing ongoing processes and conditions. For example, personnel should be encouraged to apply RCA when they see that a production process routinely yields 97% rather than 100%. Again, Figure 14-2 provides a model for this kind of decision making. Investigations into chronic loss are typically more financially beneficial to organizations than are investigations into events. Step 1 of the RCA Rt process encourages investigators to identify whether the issue is an incident or an ongoing condition.

Summary

Most companies, Beta included, do not routinely take problems to their root cause solution. In fact, it is a rare situation where this approach is common, with most businesses apparently preferring to treat symptoms. This approach results in recurring events and ongoing conditions associated with the same problems, and incurs additional production losses, costs, and risks to the business. It does not need to be this way, and Beta is like most organizations in that it would benefit substantially from promoting a root cause culture.

There are several RCA methodologies available from a number of sources. The 5 Whys appears to be the least expensive and the most easily deployed throughout an organization, providing routine improvement, so long as the management has made it clear that problems are to be taken to root cause. The more complex problems likely deserve a more sophisticated approach using the methodology similar to CEDAC for the chronic, ambiguous problems, or a methodology similar to the Apollo or RCA Rt methodology for one-time events. Both of the latter methods can be used for either situation. Having a screening tool to apply before proceeding with the more sophisticated approach also seems appropriate. Finally, for a major catastrophe, it is recommended that one of the more complex methodologies be used, along with veteran analysts to help with the analysis.

Beta intends to use the 5 Whys more extensively, including a continuum approach, to identify those times when a more comprehensive approach is needed for the more complex problems.

References

1. Fraser, A., Various Correspondence, Cheshire, England, Reliable Manufacturing Associates, circa 2002.

2. Liker, J., *The Toyota Way*, New York: McGraw Hill, 2004.

3. Moore, R., *Making Common Sense Common Practice: Models for Manufacturing Excellence*, Boston, MA and London, England: Elsevier Butterworth-Heinemann, 2004.

4. Gano, D.L., *Apollo Root Cause Analysis*, Yakima, WA: Apollonia Publications, 1999.

5. Holmes, W., *RCA Rt*, Melbourne, Victoria, Australia: SIRF Roundtables, Ltd., 2006.

Closing

15

We get brilliant results from average people managing brilliant processes. Our competitors get average results from brilliant people managing broken processes.

Hajime Ohba

Beta's Strategy

Bob Neurath is determined to assure Beta's continuing improvement and prosperity, and to slowly change the culture of the organization. Despite exceptional progress in the past few years, Beta's performance at present is simply not good enough. When their performance improvement began under his tenure they were "at the crest of a trough." There was nowhere to go but up (or out). Now they must strive for becoming the very best. A major worry is that precisely because they have made major gains, some of his management team is becoming more comfortable with their current performance, which in fact is only mediocre. Mediocrity cannot be Beta's standard of excellence.

His strategy for achieving a much higher standard is to expect excellence, and to rigorously apply the following fundamentals:

1. **Align the organization.** While great strides have been made, much better alignment is needed throughout the organization. There is still too much "silo" behavior—functional managers having greater loyalty to their personal interests than they do

to corporate interests. Superordinate goals must be created and repeatedly communicated; and he and his management team must "walk the talk." Those that don't will not stand. Bob believes that Toyota's superordinate goals are suited to Beta, although he recognizes they will take years to inculcate those into the organization's "DNA." Beta must have a long-term philosophy and strategy with the following elements:

    a. It aligns the whole organization to a common purpose that is bigger than just making money. Making money is a consequence of doing the right things.

    b. It generates value for the customer, society, and the economy.

    c. CEO pay must not be a mis-aligning issue. CEO compensation must be internally equitable and externally competitive. The CEO must put their personal interests below employee interests and corporate interests, which are first and foremost. The CEO must build trust with the employees that this is being done.

    d. Initiative overload must not be a mis-aligning issue. Initiatives must be consistent with the overall strategy and purpose of the organization, and must be introduced at a rate that the organization can absorb and effectively implement.

    e. Beta's asset management strategy will be used as a tool for further aligning marketing and manufacturing to a common set of goals.

    f. Marketing, R&D, and manufacturing must be aligned to our superordinate goals and must work together to optimize our overall business performance.

2. **Innovate.** Innovation must not be something that is only done by the R&D function. Within the organization and particularly at th4e shop floor, everyone must be constantly doing "little innovation"—looking for ways to do things better and applying rigor to the change process. This will fund the "big innovation" that R&D creates for improved products and processes. As Liker observed, innovation and standardization are two sides of the same coin. Our people are our most important asset and we will treat them as such; they are our human intellectual capital and are more valuable than our equipment.

3. **Apply leadership and management principles.** Bob will expect his managers to be leaders, and he will set the example for this. Leadership requires vision, a greater sense of purpose, watching the horizon, albeit grounded in reality. Leaders put people first, treating them with dignity, respect, and appreciation. Leaders are trustworthy, true to their word and principles, and create trust among their employees and suppliers. Leaders have a passion for excellence, set high work and ethical standards, and create a caring, disciplined, proud environment. Leaders set the example, and have the courage to support and defend their basic values and principles. Management principles will still need to be applied as appropriate; managers must administer and maintain, get things done right, manage to budgets, measure and control, hold people to standards, and constantly look for ways to improve. That said, he will be expecting that the higher in the organization his managers are, the greater the application of leadership principles that will be applied. Failure to do so will limit their future at Beta.

4. **Apply teamwork principles.** Teamwork underpenned by the superordinate goals is a must. Teams must have a clear purpose and direction; a sense of meaning aligned to corporate vision/direction; boundaries for the team's goals and self determination within those boundaries; openness and a willingness to share conflicting views; skills to achieve the goals or training needs thereto; discipline and measurement of their effectiveness or impact on the business; continuing feedback and support; flexibility to address changing needs (e.g., boundaries, training, measures); and so on.

5. **Manage change.** In this Darwinian world of capitalism, constant adaptation and improvement is a must. The compelling reasons for change are represented in Beta's superordinate goals. Beta is committed to aligning the organization to those goals. The heart of managing change, however, lies in engaging all of the employees in the change process and constantly looking for ways to do things just a bit better, and then measuring the results. People don't want to change, but they will, if they feel they have some control of the change process. As Margaret Wheatley said, "People own what they create." So we must let them create, but in

an environment where the changes are managed using a "scientific method" (e.g., Deming's Plan, Do, Check, Act [PDCA]).

6. **Improve succession management.** This issue needs particular attention at Beta. The good intention of getting young managers on an accelerated path to promotion by moving them every two years and exposing them to lots of management opportunities is apparently having a huge de-stabilizing effect on the organization. The heart of a stable business process is a reasonably stable management team with processes in place that foster standardization and innovation. Beta's policies in the future will assure much greater stability in the management team, particularly at the plants.

7. **Use the business-level Failure Modes and Effects Analysis (FMEA).** The business-level FMEA model will be used in the short term to identify major opportunities for improvement, and the appropriate tools for addressing those opportunities. However, it is expected that the leadership of each plant will engage *all employees* in constantly seeking better ways to do things, not just a few working on highly critical projects.

8. **Apply the tools.** The tools will be applied in the general priority discussed below.

Nominal Hierarchy for Application of the Strategy and Tools

After much consideration, Bob believes that Figure 15-1 represents a proper, albeit nominal, hierarchy for application of the various tools for improvement. It begins with an overall philosophy and strategy patterned after *The Toyota Way*[1]:

1. Long-term thinking is the beginning and the foundation of business excellence, even at the expense of short-term profits.

2. Understanding Beta's processes is key, including having a clear picture of where value is added and where waste is incurred, and driving out the waste.

3. Engaging and challenging Beta's people and suppliers. Being simultaneously encouraging and demanding of its people

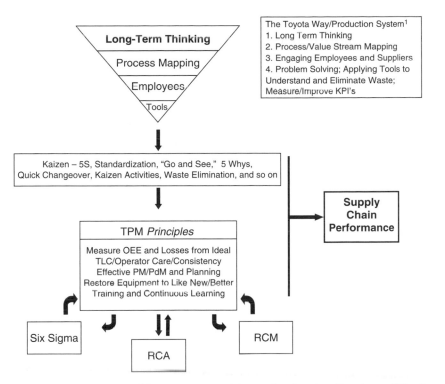

Figure 15-1 *Nominal Improvement Hierarchy for Application of Tools and Strategies.*

will be a big challenge for the leadership in getting the balance correct. Doing so is not an option.

1. When the first three steps are established, then the tools for improvement can be more readily and properly selected.

Initial priority will be given to Kaizen principles and in particular will focus on 5S, standardization of work practice, and the elimination of waste, in all its forms. Deming's principles related to PDCA, and Imai's modification of it (standardize, do, check, act), will be routinely applied in the improvement process. The 5 Whys will be used in problem solving and getting to the root cause. Quick changeover techniques will be applied to reduce production delays. Kanban techniques will also be applied, with modification made at Beta's process plants to recognize that they have inherently better "flow," and at times must "push" production. All managers must "go and see" and understand what is really happening on the shop floor, participating in kaizen events as appropriate, and then standardizing the results of those events each and every time, getting consistency in all practices.

The next priority will be given to Total Productive Manufacturing, or Total Productive Maintenance (TPM) principles, that is, measuring Overall Equipment Effectiveness (OEE) and all losses from ideal so that they can be minimized. It will include having operators apply TLC—tender loving care, doing tightening, lubricating and cleaning—monitoring equipment and process condition and having consistency of operation, particularly across shifts. It will include having best practices in Preventive Maintenance (PM), Predictive Maintenance (PdM), and planning and scheduling that are an inherent part of maintenance excellence, and will include restoring equipment to a like-new performance condition, or better. When functionality is still insufficient for business requirements, equipment and process design changes will be made. Training and continuous learning will be an ongoing part of these practices.

The next priority for the application of the various tools will be Six Sigma, Reliability-Centered Maintenance (RCM), and Root Cause Analysis (RCA) as appropriate for the more complex problems. The tool will be selected based on the nature of the problem or issue at hand. Six Sigma will typically be used for the more complex process and system-level problems where variability is far too high. As for RCA, Beta will initially apply the 5 Why methodology consistent with the Kaizen model. If that does not yield satisfactory results, then CEDAC or a more comprehensive RCA model will be used. RCM will be used to optimize PM and PdM practices and to solve more complex equipment problems. Each of the tools will require a feedback loop to foster and sustain continuous learning and improvement on the use of the tools. The business-level FMEA analysis process will be used to identify those problems and tools that deserve the highest priority.

Doing all of this will ensure excellence in supply chain management, particularly when combined with our alignment efforts between marketing, sales, and manufacturing. Customers and products will be rationalized as appropriate to the marketplace, and suppliers will be treated with respect and as an extension of Beta's business enterprise. The "supply chain" model—a supplier, Beta, and a customer—will be used as appropriate as Beta matures in its performance.

It's important to highlight that these are *nominal* priorities for the application of the tools. In any given circumstance it may be appropriate to apply RCM, Six Sigma, or RCA to a problem as part of applying Kaizen principles, or TPM, or to mix and match tools as needed under the business circumstance. However, the hierarchy and priorities shown appear to be those that will best serve Beta as it moves forward.

While Beta is still on average in its performance, it is going in the right direction and getting the right strategy, processes, and tools in place in order to align the organization to a common purpose. Having a constancy of purpose in implementing its strategy will ensure its future is bright in this highly competitive world, and that it achieves its goal of being among the best in the world.

Reference

1.   Liker, J., *The Toyota Way*, New York: McGraw-Hill, 2004.

Planning and Scheduling

> Plan the work, work the plan.
>
> *Unknown*

In his book *The Maintenance Planning and Scheduling Handbook*, Doc Palmer provides a wealth of information and knowledge regarding planning and scheduling.[1] Much of this approach to planning is summarized in the following te xt and unless otherwise noted is based on a conference presentation by Palmer.[2]

Excellence in planning and scheduling must be linked to knowledge of equipment condition, to excellence in Preventive Maintenance (PM) and basic care, and then *integrated with production planning:*

- Establish a policy for planning and scheduling and the related criteria. For example, what types of jobs will be planned? Those where two or more trades are involved? Involving contractors? Where work permits or vessel entry permits arc required? And so on. The more complex the job in terms of trades involved, complexity of equipment, safety issues, and so on, the greater the need for planning. Guidelines should be established for which jobs need planning.

- Assure planning and scheduling competency through training and practice.

- Approximately one planner is needed for every 20 maintenance technicians, considering trade skills, equipment complexity, reliability and productivity goals, and so on. Some "best practice" plants have one planner for every 12 technicians.

■   The primary goal is to improve labor productivity. For example, a typical "wrench time" is 35% spent on the job/tools. This contrasts to a "best in class" of 55%. Caution: wrench time must be balanced with other issues such as safety, complexity, commissioning, and so on.

There is extraordinary leverage in excellence in planning and scheduling. Consider the following question: where is our breakeven point for the number of planners required for doing planning and scheduling?[3]

■   Suppose we have three people at a typical 35% "wrench time," then we have the equivalent of:

$$\text{FTE} = 3 \times 35\% = 105\% \text{ (full-time equivalent people [FTE]}$$
$$\text{on the tools)}$$

■   Suppose we use one of the three for planning and scheduling to improve their productivity to 55%, then:

$$\text{FTE} = 1 \times 0\% \text{ (planner)} + 2 \times 55\% = 110\% \text{ (FTE on the tools)}$$

We would be better off having one planner for every two people in this scenario. This is not to suggest that we would actually do this, but rather to illustrate the potential productivity gains in doing proper planning and scheduling.

Let's further suppose we use one planner for 20 people to improve their productivity by 55%/35%, or 1.57. In this scenario we have actually created the equivalent of additional resources (that don't add headcount):

$$\text{FTE} = 20 \text{ people} \times 55/35 = 29.4 \text{ (or 9 extra people; @ \$50/hr}$$
$$= \$ 900K/yr)$$

We have created the equivalent of nine more people through more effective planning and scheduling. In most organizations a common complaint is that they don't have enough people to do the work. Perhaps the question that should be asked is: are we using the people we have effectively?

Planning and scheduling are not the same[2]:

- Planning is related to the job scope, sequence, requirements, resources, and so on.
- Planning relates to future work: next week, next month, and is typically *not* "**now**" work.
- Planning is not "PM" work:
    - PM prevents equipment deterioration (e.g., lubes, care, calibrations, and so on);
    - Or monitors deterioration for future action.
    - PM must be scheduled and monitored for compliance.
    - Note: PM detail must be developed at least once and modified from time to time to reflect shop floor learning and feedback.
- Scheduling is related to the priority and timing of the work.

Further, effective planning requires:

- A commitment to quality work as first priority, with a focus on PM/PdM and the work resulting from it.
- An organization separate from skilled trades.
- A focus on future work.
- Equipment files at the equipment level, including PM/PdM and bill of material needs.
- Excellence in estimating work required.
- Recognizing trades' skill (don't over-plan).

Effective scheduling requires:

- PM/PdM and jobs to be planned are identified first to allow scheduling, and integration with production schedules. Priority should be given to PM/PdM work, and the corrective work resulting from these.

- Adherence to work priority standards to avoid false emergencies.

- Knowing actual craft hours available for the scheduled period.

- Weekly meetings (suggest Wednesday) to schedule next week's jobs.

- Daily meetings to allow crew and operations supervisors to work through individual job assignments and daily schedules.

- Measuring schedule compliance.

A good plan covers:

- Work purpose and scope (visit the job site; talk with trades; gather appropriate information).

- Equipment location and identification: name, serial number, and so on.

- Parts, tools, sequencing, and time.

- Skills, resources required.

- Applicable procedures, standards, and related documents.

- Commissioning/startup requirements.

- Safety issues, lockout/tagout standards, permits.

- Housekeeping requirements on completion.

A good plan might have the following elements:

- Work order
- Procedures
- Specification sheets
- Tools list
- Parts list

- Sketches/drawings/prints
- Permits
- Manuals

Only about 25% of jobs need formal planning with detailed job steps, but may represent 50% or more of the work to be done because of its scope and complexity. At least 90% of jobs should be scheduled, and have excellent compliance to that schedule. Nonetheless, it is difficult to schedule without planning because the time estimates for the planned work support the scheduling effort. A later section in this appendix will suggest how planners can quickly plan reactive work that supports scheduling.

Planning and scheduling are also learning tools; work orders should:

- Be closed out in a timely and accurate way.
- Include fault codes for equipment analysis.
- Include comments by the trade people on the job (observations, equipment condition, and so on).
- Include time, materials, and other information needed.

A good planner has the following attributes:

- Skills and knowledge about the equipment.
- Capability for using a variety of information and data to analyze planning requirements.
- Excellence in communication and teamwork (many variables require flexibility in planning).
- Good intellect, plus discipline in work and follow-up.
- Meticulous and thorough, but flexible about dealing with high variability.

Supporting reactive maintenance vs. planned maintenance; a good planner must:

- Avoid delaying the start of urgent jobs.

- For urgent jobs, quickly scope the job, review the equipment file, and assign a skill and time estimate, but without intricate job procedures.

- Give the crafts a head start without delaying the job.

- True emergencies are not planned.

- After the emergency is over, re-visit it and develop a standard response for the library of plans.

How do we know if we're successful? We measure our performance, for example:

- PM compliance
- PdM compliance
- Schedule compliance
- Percent planned work
- Percent reactive work
- Wrench time
- Maintenance costs

Other studies have shown that in going from a reactive mode to other modes[4]:

- Doing maintenance planning almost to the exclusion of other methods, such as PdM, Asset Utilization (AU) improved only 0.5%.

- Doing PdM almost to the exclusion of other practices such as planning and scheduling, AU *declined* by 2.4%.

- Combining PdM with PM and planning and scheduling to adjust maintenance activity based on condition, AU improved 5.1%.

- Viewing planning and scheduling/CMMS work management, PM, and PdM as TOOLS to manage work, and understanding root causes, being *proactive and focusing on the elimination of defects*, AU improved 15%.

The lesson:

- Planning and scheduling by themselves will not contribute to improved AU.
- Condition monitoring (CM) by itself will not contribute to improved AU.
- It is only when we combine them that we achieve improvement; and the best improvement comes when we do both with a proactive approach for *defect elimination*.

Planning and scheduling are essential for:

- Assuring that CM and PM are performed in a timely way.
- Assuring labor productivity and equipment management.

Operational benefits of excellence in planning and scheduling:

- Crafts gain support in effective work practice.
- Supervisors gain support for effective work assignment and efficient use of resources.
- Quality and consistency of work are improved.
- Equipment histories are established.
- Impact on production is reduced.
- Planning spares/resources is improved.

Business benefits of excellence in planning and scheduling:

- Improved productivity
- Improved production output
- Improved safety
- Lower costs
- Higher profits

References

1. Palmer, R., *The Maintenance Planning and Scheduling Handbook*, 2nd edition, New York: McGraw Hill, 2005.

2. Palmer, R., *Making Planning Successful to Improve Productivity*, Sixth Machinery Reliability Conference, Phoenix, AZ, April 2001.

3. Baldridge, R., Various discussions, 2005.

4. Ledet, W., *The Manufacturing Game*, Annual TQM Conference, Kingwood, TX: November 1994.

Performance Measurement[1]

> If you don't measure it, you don't manage it.
>
> *Joseph Juran*

If you do measure it, you will manage it, and it will improve. With this in mind, it is critical that the measures selected be the correct ones, and that they be effectively communicated. The basic tools and strategies are covered in the main body of the book. However, we must make sure we measure those things that are supportive of our use of these tools and of our corporate objectives. We need leading indicators: are we doing the right things? And, we need lagging indicators: are we getting the right results? Further, management must provide the systems for all measures to be easily made to facilitate data collection, reporting, and action.

Leading indicators tend to be shop floor oriented, for example:

- Operators—leading indicators:
  - Operator care PM conformance
  - Process conformance/non-conformance
  - Equipment downtime/delay times/life
  - Housekeeping conformance
  - First pass-first quality yield
  - Other measures that operators directly control or influence

- Maintenance skilled trades—leading indicators:
  - % PM compliance to schedule
  - Mean time between repairs
  - Seal life: number of seals per month
  - Bearing life: number of bearings per month
  - Lube compliance
  - Number of leaks per month
  - Other measures that technicians directly control or influence

If we're doing the right things, then we should see substantial improvement in our lagging indicators, which tend to be management-oriented, but also tend to be more like looking in the rear-view mirror—lagging indicators show us where we've been, not where we're going. Leading indicators show us where we're going. Examples of lagging indicators include:

- Executive suite—lagging indicators:
  - Return on net assets
  - Market share
  - Sales growth
  - Earnings growth
  - Safety and environmental Performance

- Plant management—lagging indicators:
  - Unit cost of production
  - Asset utilization (AU) or overall equipment effectiveness (OEE)
  - On-time delivery
  - Safety and environmental performance
  - Maintenance cost as a percent of asset replacement value
  - Inventory turns
  - Return on net assets
  - Reactive maintenance levels

At one of Beta's principal divisions, the following were chosen as key lagging measures:

Corporate strategic:	Earnings per share/growth
	Return on net assets
	Safety (injury rate)
	Market Share Growth
Plant operations:	AU/OEE
	Return on net assets
	Unit cost of product
	On-time, in-full rate
	Inventory turns on product and stores
	Safety (injury rate)
	Percent operations downtime
	Operations cost as a percent of production costs
Plant maintenance:	Maintenance cost as percent of asset replacement value
	Maintenance cost as percent of production costs
	Maintenance downtime production losses
	Percent reactive work
	Overtime rate
	Safety (injury rate)
	Average equipment life

Additional possible measurements are provided in Tables B-1, B-2, and B-3 and in reference 2 for your consideration when establishing your key performance indicators. Select the three to five that are most important to the functional group and display those.

It's important that whatever measures you select that they "cascade" from the top of the company to the shop floor. Along the way, they must support upward to the next highest level and across departments;

Table B-1 *Example Production Measures*

Total production throughput (pounds, units, $, so on, total and
by-product line)
AU (% of theoretical capacity)
OEE (availability × efficiency × quality)
Conformance to plan (%)
Total costs ($)
Cost per unit of product ($/unit)
Maintenance cost per unit of product ($/unit)
Equipment availability by line/unit/component (%)
MTBF of production equipment (days)
Scrap rate ($ or units)
Re-work rate ($, units, number)
Quality rate ($ or units)
Finished goods inventory turns
Inventory "backlog" (weeks of inventory available)
Work in process (non-finished goods inventory, $, or units)
Overtime rate (% of $ or hours)
Personnel attrition rate (staff turnover in %/yr)
Product mix ratios (x % of product lines make up y% of sales)
Energy consumed ($ or units, e.g., KWH, BTU, and so on)
Utilities consumed (e.g., water, wastewater, nitrogen gas, distilled water,
    and so on)
On-time deliveries (%)
Returns ($, units)
Set up times (by product line)
Cycle times (plant, machine, product)
First-pass yield (product, plant)
Process efficiency(s)
OSHA injury rates (recordables, lost time per 200K hours)
Process capability (to hold specification, quality)
Training (time, certifications, and so on)
Productivity ($ per person, $ per asset, and so on)
Scheduled production and/or downtime
PM work by operators (%)

Table B-2 *Example Maintenance Measures*

Equipment availability (% of time)
OEE (availability × efficiency × quality)
Equipment reliability (MTBF, life, and so on)
Planned downtime (%, days)
Unplanned downtime (%, days)
Reactive work order Rate: emergency, run-to-fail, breakdown, and so on
  (%, %hrs, %$)
Product quality (especially in equipment reliability areas: rolling mills,
  machine tool, and so on)
Maintenance cost (total and $ per unit of product)
Overtime rate (% of $ or hours)
Personnel attrition rate (staff turnover in %/yr)
Re-work rate ($, units, number)
Spare parts/MRO inventory turns
Overtime rate (% of $ or hours)
Personnel attrition rate (staff turnover in %/yr)
OSHA injury rates (recordables, lost time per 200K hours)
Process capability (to hold specification, quality)
Training (time, certifications, and so on)
Productivity ($ per person, $ per asset, and so on)
Scheduled production and/or downtime
Bearing life: actual vs. design life
Mean time between failure, average life
Mean time to repair, including commissioning

Planned and scheduled work/total work (%)
PM/work order schedule compliance (% on schedule)
Hours covered by work orders (%)
PM work by operators (%)

PMs per month
Cost of PMs per month
"Wrench" time (%)

Average vibration levels (overall, balance, align, and so on)
Average lube contamination levels
Schedule compliance for condition monitoring
Predictive maintenance (PdM) effectiveness (accuracy of PdM by
  technology)

Table B-3 *Other Related Measures*

Other corporate measures:
Return on assets
Return on equity
Return on invested capital
P:E ratio
Percent profit

Other maintenance measures:
Mechanics, electricians, and so on, per
Support person
First-line supervisor
   Planner
   Maintenance engineer
   Total site staff
Total number of crafts
Training hours per craft
Maintenance cost/plant replacement value
Plant replacement value $ per mechanic
Stores $/plant replacement value $
Stores service level: stock out %
   Critical spares
   Normal spares
Contractor $/total maintenance $
Maintenance $/total sales $
Maintenance $/value added $ (excludes raw material costs)

Industry specific measures:
Refining: dollars per equivalent distillate capacity
Automotive: hours per automobile
Electric power: equivalent forced or unplanned outage rate

---

they must not be conflicting. For example, driving marketing and sales to sell high-volume products with low margins, while improving sales may not improve return on assets. Or, having production delay needed maintenance to focus on production in the short term, may have a detrimental effect on maintenance costs, and longer term on plant capability. There should also be a few higher order or "superordinate" goals and measures that everyone must support, no matter where they are in the organization. Finally, remember that nothing changes until the shop floor does things differently, so it's really important that we have measures at the shop floor level that facilitate their doing a better job and provide timely feedback on their continuing progress.

Return on Net Assets (RoNA) or Return on Replacement Value (RoRV)

Many corporations use RoNA or some comparable measure as part of determining operational and financial performance. While this is an excellent measure, it may be more appropriate to measure specific plants on the basis of Return on Replacement Value. In those cases where a plant has been depreciated to where it has a low book value, RoNA may not truly reflect the plant's performance, as compared to a new plant. It is more appropriate to put all plants on a normalized basis and to truly judge the plant's operational performance.

Measure for Weaknesses

One of Beta's divisions subscribed to the philosophy that measurements should also expose their weaknesses, not just their strengths. There is a natural tendency in almost everyone to want to look good. Hence, if measurements are made, we want to score high, and thus look good. However, this tendency can at times lead to a false sense of security. For example, at one Beta plant the plant manager reported an OEE of 98%. However, on closer review, this was being measured against an old production rate that was established before several de-bottlenecking efforts. After considering this and factoring in all the other issues, the plant had an OEE closer to 70%.

Benchmarking against the best companies will help minimize the tendency to use measures that assure looking good. Making sure that you're setting very high standards for your measurements will also help in this. As stated, measurements should expose your weaknesses, not just your strengths. Only then can these weaknesses be mitigated or converted into strengths.

Beta's Instrument Division

The new vice president of manufacturing for Beta's small start-up instrumentation division questioned the plant manager about the quality of the plant's new products. The plant manager assured the vice president that the products' quality was good, even very good. When asked for the measures of quality, however, the plant manager admitted that their focus was on finished product quality, since there was a strong commitment to customer satisfaction; new products tended to get "buried" in the total statistics, since the more mature products tended to represent the greatest volume. Soon thereafter, measurements were intensified on all products, and the vice president focused on two key measures: (1) for internal purposes—first-pass yield; and (2) for external purposes—DOAs, or equipment that was "dead on arrival," or failed to meet any critical function within the first 30 days of shipment. Supporting measures were also implemented, but these two were considered key measure of success.

After collecting the information for several months, it turned out that quality was not good in either of the key measures for new product success. DOAs were >3%, or 3 of 100 customers could not use the product because of defects that rendered a key function useless. This was particularly painful, since it was their belief that dissatisfied customers will tell 10 to 20 people of their bad experience and, therefore, the potential existed for some 30% to 60% of potential customers not to view the company favorably. It also turned out that first-pass yield was only about 66%, meaning that effective costs for manufacturing were 50% (1.0/0.66) above ideal.

The good news was that critical data was now being measured, and it could be managed toward improvement. Clearly the next step was to examine the causes for DOAs and low first-pass yields. Upon investigation, causes were determined to be mechanical failures, component failures, assembly failures, software failures, and so on. A Pareto analysis was performed, which identified the principal causes of the failures and provided a basis for establishing priorities for solving them.

A team was formed to evaluate, on a prioritized basis, the cause of the failures, and then to quickly implement change for eliminating all failures. For example:

1. Each instrument was fully powered for a period of time sufficient to minimize startup and early life failures, and a series of tests were developed and documented in a procedure (check-off) to assure mechanical and electrical performance. Non-conformance judgments were placed in the hands of each technician who had the authority to reject any equipment under test. Technicians were also encouraged to work directly with manufacturing and engineering to improve processes to eliminate failures.

2. Review of the Pareto analysis and a root cause failure analysis indicated vendor quality problems were at the source of many problems. Components incurring a higher rate of failure were reviewed directly with the vendor to develop corrective action. A receipt inspection program was also implemented to assure quality products were received.

3. Mechanical defects were particularly troublesome. In reviewing the data, specific patterns emerged as to failure modes. Design modifications were implemented. Procedures were re-written (engineering and manufacturing) to eliminate failures. Each instrument was put through a mechanical shake test to simulate several severe cycles, validating the mechanical reliability of the equipment.

4. Software was rewritten and test procedures were made more stringent to assure system functionality.

5. A customer service line was established to provide details of failures, failure modes, and effects, and these were entered into a database for use in future designs and design modifications.

The measurement and analysis process continued forward (continuous improvement). Over the course of two years, first-pass yield increased to over 85% (a 21% cost reduction), and the DOA rate dropped to 0.15%, a 20-fold improvement. While neither of these was considered world class, the trend was clearly in the right direction, and the division expected that in time world-class performance could be achieved.

When new products were introduced, these measures would sometimes show a deterioration in performance, but the impact was immediately recognized, analyzed, and corrective action taken to improve the quality, both of the product and of the processes. Customers received higher quality products, and first-pass yield increased (costs were reduced).

Using this success as a model, performance measures were implemented throughout the plant with comparable results. For example, in customer support, measures of customer satisfaction were implemented and routinely surveyed. Customer support staff were strongly encouraged to use their own judgment and take the steps they deemed necessary to satisfy customer needs. Likewise they were strongly encouraged to seek help from and give input to the engineering, manufacturing, and sales departments. If in their judgment, they felt the problem needed senior management attention, they were encouraged to immediately contact the president or other officer of the company to seek immediate resolution. This was rarely necessary, but demonstrated corporate commitment to customer satisfaction. Over time, the staff became outstanding and handled essentially all situations relating to customer satisfaction, achieving a 93% overall approval rating.

Other departments, likewise, showed substantial improvement in their performance—accounting improved cash flow from 53 days aged receivables to 45 days aged receivables, cut billing errors in half, provided month end reports within five days of the end of the month, and so on. Engineering had fewer design defects, improved design cycles, and so on. Some additional lessons learned:

1. Simply measuring data is not sufficient to assure success. Performance measures must be mutually supportive at every level within the company and across company departmental lines. Measures that encourage an attitude of "I win if I do this" only alienate other departments and individuals. Measures that focus on the customer winning and, therefore, the company winning will assure a greater degree of success. They must **not**, as the Management by Objectives (MBO) strategy tended to do in the 1970s, create narrowly focused interests and the loss of teamwork. The least desirable measure is one which results in a reduction or loss of teamwork. As such, measurements must be integrated into a hierarchy that is supportive of overall corporate objectives and does not conflict with other departmental measures. Measurements must be mutually supportive between departments.

2. Measures must be displayed clearly, openly, and unambiguously. This provides an emphasis of the importance to the company, to the customer, and, therefore, to the individual. It also provides employees with feedback on individual, department, and company performance, and allows employees to adjust their activities to improve their performance according to the measures. Processes and methods can then be revised to assure continuing improvement.

   At Beta's Watergap plant, key performance measures related to output, quality, and unit cost were boldly and clearly displayed for all to see. However, on closer review, the data was nearly three months old. When questioned about this, the production manager replied that "We'd had a couple of bad months and we didn't want to concern people." The philosophy that should be taken in companies is that when you have bad news, get it out right away. If you don't, the rumor mill will capture it and amplify it. More importantly, the employees could take the view that management doesn't trust the people well enough to be honest with them. Of course, we all want to hear good news, but the real test of a company is often how it manages bad news, and particularly the plan of action to assure employees that the problems will be resolved.

3. For a given functional unit, limit the number of measures displayed to five. More than this, and the measures get cluttered and confuse any message of importance trying to be conveyed. People can only respond to a few measurements effectively. Any particular manager may want to track other measures which support the key three to five, but displaying them will likely detract from the goal of using measurements to reinforce or modify behavior.

Closing

Let's close with some advice from Jack and Suzy Welch. They recommend three key performance indicators, irrespective of the size of the company[3]:

Employee engagement

Customer satisfaction

Cash flow

Employee engagement has less to do with trivial things, like office size and cafeteria food, and more to do with whether or not the leadership is "walking the talk," and giving people a sense of direction and meaning aligned to the corporation's overall strategy and objectives. This, along with employee engagement is also a key principle at Toyota.

Customer satisfaction requires being with customers and understanding their needs and level of satisfaction with the company's performance in meeting those needs. Customer satisfaction as determined by quality, cost, and delivery, is likewise, a key principle at Toyota. Both of these are more subjective and difficult to measure, but can be accomplished through well-designed surveys and routine communication with employees and customers.

The last, cash flow, never hides things. Generally accepted accounting principles almost always have assumptions regarding the numbers (e.g., how you handle revenue accrual, depreciation and amortization, warranty liabilities, and so on). There is flexibility in all of these accounting treatments. However, all of a company's financials ultimately end up as cash, no matter the accounting treatment. Cash will always tell you the true condition of the company.

If you measure it, you will manage it, and it will improve.

References

1. Moore, R., *Making Common Sense Common Practice: Models for Manufacturing Excellence*, Boston, MA and London, England: Elsevier Butterworth-Heinemann, 2004.

2. Haskell, B., *Performance Measurement for World Class Manufacturing*, Cambridge, MA, Productivity Press, 1991.

3. Welch, J., and Welch, S., *Ideas The Welch Way – How Healthy is Your Company?*, Business Week, May 8, 2006.

Listing of Commonly Used Improvement Tools and Terms

C

So many tools, so little time...

In previous chapters several of the more common and popular improvement tools and strategies were reviewed. It was also indicated that there are likely 100 or more of these tools available, so clearly not all of them were covered. Some of those not reviewed, while not as popular or common as those we've considered, deserve some mention. Those are discussed below. If your favorite tool has been omitted, that's not to say that it's not useful or important, but rather that it may not be known to the author. A listing of these additional tools is provided in Table C-1 and comes from SIRF-Roundtables, a networking organization in Melbourne, Victoria, Australia.[1] These include many of those discussed in previous chapters.

Table C-1 *Listing of Improvement Tools*

1. ABC Inventory Control
2. Activity-Based Costing (or Accounting)
3. Aggregate Planning
4. Agile and Quick Response
5. Andon
6. Automation
7. Benchmarking
8. Capacity Planning

Table C-1   *Listing of Improvement Tools*   *(Continued)*

9. Capability Index
10. Cellular Manufacturing
11. Concurrent Engineering
12. Design for Manufacturing and Assembly
13. Design for Maintenance
14. Design of Experiments
15. Enterprise Resource Planning
16. Factory Layout
17. Failure Modes and Effects Analysis (FMEA)
18. High Performance Teams
19. Inventory Reduction
20. Job Design
21. Just in Time
22. Kaizen
23. Kanban
24. Leadership/Supervisor Development
25. Lean Manufacturing
26. Machine Efficiency
27. Nationally Accredited Training
28. Networking
29. New Equipment
30. Operator Maintenance
31. Open Book Management
32. Overall Equipment Effectiveness (OEE)
33. People Development
34. Planned Maintenance
35. Poka Yoke
36. Postponement
37. Predictive Maintenance (PdM)
38. Project Management
39. Quality Function Deployment (QFD)
40. Quality System and Accreditation
41. Re-Engineering
42. Reliability-Centered Maintenance (RCM)
43. Risk Analysis
44. Root Cause Analysis (RCA)
45. Sales and Operations Planning
46. Setup Reduction
47. Seven Tools of Quality
48. Single Minute Exchange of Die (SMED)
49. Skills Analysis
50. Skills Matrix

Table C-1 *Listing of Improvement Tools* *(Continued)*

51. Software Systems
52. Statistical Forecasting
53. Statistical Process Control (SPC)
54. Statistical Testing
55. Suggestion Schemes
56. Takt Time and Line Balancing
57. Target Costing
58. Team Development
59. Theory of Constraints (TOC)
60. Time Studies/Work Method Studies
61. Total Productive Maintenance (TPM)
62. Toyota Production System (TPS)
63. Value Analysis
64. Value Engineering
65. Value Stream Mapping
66. Visual Manufacturing
67. Waste Reduction

Several of these are covered in some detail in various chapters of this book. Others that come to mind, but are not on this list include:

68. The Manufacturing Game®
69. Kepner Tregoe®
70. 5S
71. 5M
72. 5 Whys
73. Risk-Based Maintenance
74. Proactive Maintenance
75. Mission Directed Work Teams

No doubt there are still many others. A brief summary of each of these is provided below.[1]

1. ABC Inventory Control

Appropriate inventory control processes are concerned with matching the effort expended in managing inventory to the value of the inventory. The ABC classification system is used to rank inventory units according to their value.

Classification	% Dollar Value	Approximate Percentage of Items
A	80%	20%
B	15%	30%
C	5%	50%

This system is used to assist in determining the level of effort that should be applied to managing the various inventory units. Appropriate forecasting and inventory monitoring systems can be developed for each class.

2.   Activity-Based Costing (or Accounting)

Activity-based costing is a technique used to determine the cost of servicing particular customers or producing products by understanding the activities involved in so doing. Costing systems in many companies are designed to assist in the development of financial reports, not to aid in making operational decisions. Activity-based costing addresses this gap by providing the information that allows informed decisions to be made.

3.   Aggregate Planning

Aggregate planning is a technique for planning production with requirements at a high enough level to assess if it is consistent with the company's strategic goals and objectives without getting bogged down in detail. It involves aggregating demand into product families based on production requirements (e.g., labor, materials, routing and so on), aggregating labor (e.g., into units based on capability or expertise), and aggregating time into periods (e.g., weeks or months) to simplify the planning process. The plan can be assessed for general do-ability and as to whether it meets the company's financial goals.

4.   Agile and Quick Response

Agile manufacturers recognize the volatility of change, and put mechanisms in place to deal with it. They move from being manufacturing-driven to customer-driven, and they also realize customers won't pay a premium for quality—it's assumed. Agile manufacturers partner with customers, suppliers, and competitors (cooperate and compete), and understand that the soft side of business (trust, empowered teams, risk taking, reward, and recognition) drive the entire process. In an agile environment, information is the primary enabling resource. Firms must know their customers, products, and competitors.

5.   Andon

This is a simple visual system consisting of a visible light or sign that shows the state of an operation. Its purpose is to quickly inform the appropriate people when there is a problem so they can attend to it. When lights are used, they are normally coded:

Green—OK
Orange—Problem
Red—Breakdown

Other lights are sometimes used to denote material shortages.

6.   Automation

The use of integrated manufacturing systems (machinery) that seek to lower cost per unit and improve performance, quality, and/or responsiveness. These results are generally achieved by using machinery of sufficient sophistication to significantly reduce or eliminate associated production labor.

7.   Benchmarking

Benchmarking is aimed at rapidly improving the efficiency and effectiveness of operations by understanding how other companies are performing the same or similar operations and adopting practices that prove superior. Briefly, the technique involves:

1. determining which process or operation needs improving
2. developing a clear understanding of the process

3. seeking benchmarking partners, preferably ones that are known for exhibiting best practice in the area under study;

4. developing a questionnaire and conducting a site visit;

5. analyzing the results of the questionnaire and site visit; and

6. developing an implementation plan for transferring the practice.

8. Capacity Planning

The purpose of capacity planning is to determine if a plant has the capacity to meet current and future production demands. It requires knowledge of the degree of capacity utilization (i.e., the degree to which plant and equipment are currently being used), the peak capacity (i.e., the maximum capacity under ideal condition), the effective capacity (i.e., the maximum sustainable output), and the capacity requirements of products. Production plans are assessed using this information to identify bottlenecks and determine whether requirements can be meet with available equipment.

9. Capability Index

This is a means of assessing whether a machine or a process has the ability to consistently produce material within the required specification. It involves calculating the mean and standard deviation of the critical characteristics produced via the process and then comparing these to the specification for these characteristics. If the process is under statistical control (i.e., it is not subject to special causes of variation) and its capability index is >1, then the process will need to be changed to consistently produce output with the required quality.

10. Cellular Manufacturing

Cellular manufacturing—an approach in which manufacturing work centers (cells) have the total capabilities needed to produce an item or group of similar items—contrasts to establishing work centers on the basis of similar equipment or capabilities, in which case items must move among multiple work centers before they are completed.

11. Concurrent Engineering

The traditional way of designing a product involves a sequential series of activities where the results of each stage in the development are passed to the next group of people. As a result, rework is often required as a design feature developed at an early stage and is later discovered to be unsuitable for manufacturing, logistics, or marketing. Concurrent or simultaneous engineering (or to use the most recent buzzwords, Rapid Product/Process Realization ([RPPR]) is a process that involves the participation of all the stakeholders of the product development process at critical stages in the design process. This is to ensure that logistical and manufacturing constraints are considered in the early concept development and design stages. Concurrent engineering can lead to significant reductions in the product development lead-time.

12. Design for Manufacturing and Assembly

Design for manufacturing and assembly refers to a number of techniques for ensuring that the manufacturability of a product is considered during the design process. Factors to be considered include delaying of customization of the product, minimizing the number of components (both within the product and across the product family), degree of modularity, use of symmetry to allow easier handling, use of grooves and guides to assist in assembly, and use of common tooling.

13. Design for Maintenance

Design for maintenance is the fourth component of a TPM program. It involves engineers and operators working together to design equipment that is easy to maintain. This may involve reducing the numbers of components in machinery that need to be maintained or improving access to vital areas likely to experience wear. Techniques like brainstorming, cause and effect, and root cause analysis are often applied to identify the major causes of failure to lead the redesign process.

14. Design of Experiments

This is a technique that can be applied to identify the major cause of a problem when the cause is hidden among a large number of interacting variables. This technique significantly reduces the total number of experiments that need to be performed to account for the effects of each variable on each other.

15.   Enterprise Resource Planning

This is the practice of consolidating an enterprise's planning, manufacturing, sales, and marketing efforts into one management software system.

16.   Factory Layout

These are a series of techniques to optimize the layout of machinery within the factory. They include techniques such as group technology and cellular manufacturing, which involve placing machines into groups based on the frequency and sequence of processing steps required for particular product families, and spaghetti diagrams for determining the distance and route through which products or people need to move during normal activities.

17.   Failure Modes and Effects Analysis (FMEA)

See number 43 on Risk Analysis.

18.   High Performance Teams

The development of high performance teams is a goal rather than a technique. The aim is to transfer the determination of as many operational decisions as possible to the people with the required knowledge, i.e., shop floor personnel, to improve the speed and reliability of these decisions. A number of features are required to support such an environment, including the development of clear performance indicators, appropriate training and job design, and appropriate reward systems.

19.   Inventory Reduction

A program designed to reduce raw material, work in progress, and/or finished goods holdings. Generally the reduction is achieved by moving to smaller batch sizes or even single piece flow; it could also be achieved by improving responsiveness by reducing cycle times or by better forecasting demand.

20. Job Design

Job design involves assessing jobs with respect to both their physical and psychological characteristics. On the physical side, aspects such as ergonomics, repetitiveness of movement, and effort required are analyzed to determine if they have a deleterious affect on performance. Likewise, psychological aspects such as boredom or ability to influence work outcomes are determined and efforts are directed towards improving these aspects.

21. Just In Time

This is a manufacturing method where downstream operations pull required parts needed from upstream operations at the required time. Implementing just in time requires most features of Lean Manufacturing.

22. Kaizen

Process improvement that involves a series of continual improvements over time. These improvements may take the form of a process innovation (event) or small incremental improvements.

23. Kanban

Kanban is a system using signals such as cards, containers, or painted spaces on the floor, which are linked to the customers' demand for communicating when to process material. The purpose is to make only what the customer needs by using the customer's demand as a signal to pull production through the plant, rather than produce according to a plan or forecast. It is mainly used in a classic mass production environment with few product variations and leveled demand. A more recent variation known as POLCA (Paired, Overlapping, Loops of Cards with Authorization) is applicable in a job shop environment where each job can be unique.

24. Leadership/Supervisory Development

This is a program that seeks to broaden skills and or attempts to change behavior or perceptions of people who work in leadership or supervisory roles. The training generally deals with creating a positive and beneficial interaction between supervisor and subordinate but also details disciplinary procedures.

25. Lean Manufacturing

Lean manufacturing is a philosophy, drawing heavily on the Toyota Production System, focused on making value flow to the customer in the most effective and efficient manner by producing goods (or services) using the minimum amounts of cost, time, space, and resources. It employs a wide range of techniques designed to identify and eliminate all forms of waste or non-value adding activities from the manufacturing process as it moves towards its goal of achieving a batch size of one. A company adopting lean principles often achieve significant reductions (>50%) in costs, inventory levels, and lead-time.

26. Machine Efficiency

The aim of techniques under this heading is to improve asset utilization by ensuring that machines are available for use when they are required. The techniques are directed toward elimination of what are known as the six big losses. These losses are:

Breakdown
Set-up and adjustment
Minor stoppages
Reduced speed
Quality defects
Startup and shutdown

27. Nationally Accredited Training

Nationally accredited training has standardized outcomes and competencies and is nationally recognized. The Frontline Management program is an excellent example and is relevant to manufacturing.

28. Networking

Networking is an association of companies who discuss operational strategies and issues, share ideas, and learn from their peers. The interactions between companies facilitates and accelerates learning.

29. New Equipment

The business sees improvement being achieved through the purchase and installation of new equipment that may be to improve throughput, quality, flexibility, safety, and/or reliability.

30. Operator Maintenance

Operator maintenance is maintenance carried out on a machine by the operator when it is safe, economical, legal, and logical. The theory is that the operator runs the machine and through constant observation and regular inspection can detect possible problems before they develop into more substantial problems that can jeopardize the reliability of machinery and be more costly to prepare.

31. Open Book Management

Open book management is a philosophy based on making the companies' results available to all employees. The premise is that employees who are aware of the effects of their efforts on the performance of the company will act for the betterment of the company. This approach is often associated with visual management where the cost of production and company performance are openly displayed above the work areas. Usually, open book management is introduced with some basic training to employees on how to read and understand profit statements and balance sheets.

32.   Overall Equipment Effectiveness (OEE)

OEE is a system for monitoring the effective use of equipment. The OEE measure considers product quality, equipment's designed production rate, planned production time, and equipment breakdown in calculating a metric for equipment effectiveness. Improvements in quality, equipment up-time, and production rate will all increase the OEE figure.

33.   People Development

Programs that seek to broaden skills, attempt to change behavior, or change perceptions of people working in an organization.

34.   Planned Maintenance

Planned maintenance is concerned with understanding the behavior of machines to determine when maintenance needs to be conducted. An understanding of the optimal balance between the cost of preventive maintenance and cost of corrective maintenance drives the development of maintenance programs. Calculations of mean-time-to-failure of machines, or individual components on machines, are used to identify the appropriate level of maintenance. Identifying and monitoring signs that indicate deviations from normal machine behavior is actively pursued.

35.   Poka Yoke

Poka yoke involves developing failsafe devices to prevent further processing of products with quality problems. This can include designing products that can only be assembled in a particular manner, designing additions to machines to prevent incorrect processing of material, or developing methods to ensure that the correct number of units is available for assembly.

36.   Postponement

Postponement is a simple technique that is useful in an environment where demand for customized products is highly variable and the production lead-time is so long that a make-to-stock (make to forecast) production model must be used. It allows products to be planned and produced to a point before completion and prior to customization. This technique allows the common components that compose the final product to be forecast with a much higher level of accuracy than is possible for the final product. It also reduces the lead time (i.e., the time from postponed intermediate to finished product could be significantly less than the time from raw material to finished product) providing opportunities to reduce inventory and increase responsiveness. Use of this technique may require considerable redesign of the product to delay significant customization to the later stages of processing.

37. Predictive Maintenance (PdM)

Predictive maintenance (condition-based) is characterized by practices that are based on equipment condition. Examples include charging a bearing before it fails based on vibration analysis, changing lubricant based on an oil analysis showing excess wear particles, replacing steam traps based on ultrasonic analysis, and so on.

38. Project Management

Project management consists of a number of techniques that are especially useful for manufacturing in an environment where the process time involved is long (months) and each product has a high degree of uniqueness. The specific techniques include Gantt Charts (graphs that pictorially represent the duration and interdependence of tasks), program evaluation and review methods (PERT), and critical path method (CPM) that assist in scheduling and prioritizing activities.

39. Quality Function Deployment (QFD)

QFD is a technique used to capture the "voice of the customer" and translate it into the required functionality and attributes of the product. The requirements of the customer (both spoken and unspoken) are identified and ranked according to importance. The requirements are then matched with the means of achieving them, usually in the form of engineering specifications. The specifications can also be compared with each other to determine if there is a positive or negative interaction. Where there is a negative interaction, the means can be cross-referenced to the customer's requirements to determine which aspect is the most important.

The technique also allows for a comparison against both the ability of competitors to meet the customer's wants and the ability of competitor's to meet the technical requirements. Several iterations of QFD are usually performed. First, we should turn customer requirements into design requirements, and then translate the design requirements into part quality characteristics. The next step is the translation of quality characteristics into key process operations needed to provide them, and finally develop the production requirements based on the key process operations. The complete process ensures that all decisions stem from the initial customer's requirements.

40.   Quality System and Accreditation

Implement a formal quality system for third-party accreditation.

41.   Re-Engineering

Re-engineering involves determining what the organization wishes to achieve from each of its processes and then design the steps that will achieve that goal in the most expeditious manner. Frequently, processes within an organization have evolved to their current state as the demands and requirements placed upon them have changed. As a result, many processes are not optimized toward achieving their current purpose. Re-engineering allows processes to be redesigned to remove waste and non-value adding steps.

42.   Reliability-Centered Maintenance (RCM)

RCM is a maintenance methodology based on identifying the possible modes of failure of an item of machinery then developing countermeasures for each of the failure modes.

43.   Risk Analysis

Risk analysis is concerned with identifying and then eliminating or reducing the risk associated with a new product or design. A tool that can be applied to this end is Failure Modes and Effects Analysis (FMEA). In an FMEA analysis all the things that could potentially go wrong are identified (either through brainstorming or by using a more formal process based on lists of potential failures). The impact or severity of the failure is then estimated and recorded (e.g., ranked between 1 and 10). A list of potential causes for each identified failure mode is generated and both the probability that each event will take place is estimated and the ease of detection of a problem is estimated (e.g., again ranked between 1 and 10). When the product of severity and probability of occurrence is above a set threshold, a design review must be conducted to reduce or remove the risk.

44. Root Cause Analysis (RCA)

RCA is aimed at addressing the root cause of problems. Many of the things we actually consider to be problems (i.e., high inventory levels and poor service performance, are actually just symptoms of a deeper problem). Addressing the symptoms will, at best, only provide temporary relief. RCA is a means of drilling down to the real problem(s), which may be several steps removed from the symptoms being witnessed. Resolving the cause of the ultimate problem should result in the removal of all the symptoms.

45. Sales and Operations Planning

Sales and operations planning is a process aimed at developing a single production plan that has the agreement of (most commonly) sales, marketing, operations, R&D, and finance. It allows the business to operate off a single set of numbers. The process has several stages, prior to the official sales and operations planning meeting, in which representatives from the various functions identify and attempt to resolve any inconsistencies between the customers demand and the organizations' ability to supply. The final meeting is to confirm the production plan and address any longer-term issues.

46. Set up Reduction

See number 48 on Single Minute Exchange of Die, also called Quick Changeover.

47.   Seven Tools of Quality

The seven tools of quality refer to the group of quality tools initially associated with TQM. The tools are used to uncover the causes of quality problems, develop and rank solutions, and monitor the results of improvement efforts.

The seven tools are:

1.   The histogram
2.   The scatter chart
3.   The Pareto chart
4.   The cause and effect diagram
5.   Check sheets
6.   Control charts
7.   Various graphs

48.   Single Minute Exchange of Die (SMED)

SMED represents the goal of being able to perform a changeover or set-up within a minute. It involves separating aspects of the set-up into those that can be done while the machine is operating and those that require the machine to stop, and then completing as much of the changeover as possible before the previous operation has finished. Standardization of tool-jigs and use of quick-change screws, clips, or magnets can be used to gain significant time improvements. Two newer goals for rapid changeover, Single-Touch Exchange of Die (STED) and No-Touch Exchange of Die (NTED), extend these principles and involve greater use of mechanization.

49.   Skills Analysis

A skills analysis involves understanding all the tasks involved in conducting an operation or process, and determining the requisite skills for each task. This feeds into a gap analysis to determine if the current skills base is sufficient to meet the organizations' needs. A training plan can then be developed to address any current deficiencies in skills, and to ensure that planned future requirements can be met.

50. Skills Matrix

A skills matrix is a visual management technique. It consists of a chart with employees' names along one axis and required skill sets along the other axis. As employees acquire skills, some form of indicator is placed at the intersection of the name and the skill set. Various indicators (usually different colors) can be used to represent different levels of skill attainment such as beginner, competent, or able to instruct. A skills matrix can serve both as a means for management to quickly allocate appropriately trained employees to tasks, and as a device for motivating personnel to increase their skill level.

51. Software Systems

This focuses on improving the performance of the business by utilizing software that may speed the flow of information, streamline decision making, enable better control, and/or reduce costs to the business. Software may be specific to a particular business function or technical application or it could be a fully integrated business-wide application.

52. Statistical Forecasting

There is a wide range of statistical techniques for interrogating historic demand data to identify predictable trends. These can range in complexity from simple moving averages to more complex techniques that take into account seasonal factors and growth, to methods that use regression analysis to identify causal factors. A measure of the accuracy of forecasting can be used to determine the level of safety stock required to achieve a set level of customer service in a make-to-stock production environment.

53. Statistical Process Control (SPC)

SPC is a technique for monitoring the progress of a process so as to detect when there has been a shift in the underlying nature of the process: the technique can be used to determine if a process is under control (i.e., if the behavior is only subject to common causes) and what effect improvement initiatives are having.

54. Statistical Testing

There are a large number of techniques available for testing the results of changes to processes or for testing the statistical significance of observed differences in performance (e.g., testing whether the apparent difference in the performance of two shifts is statistically significant). Methods include analysis of variance and various hypothesis-testing techniques (e.g., t-test or chi-squared). These techniques are crucial for an organization aiming at attaining Six Sigma quality levels (i.e., approximately three defects per million).

55. Suggestion Schemes

This method seeks improvement ideas from the workforce. The numbers of suggestions per employee is occasionally used as a benchmark of employee morale. Sometimes the successful implementation of a suggestion may result in a bonus to the employee who made the suggestion.

56. Takt Time and Line Balancing

Takt time and line balancing are two of the cornerstones of Lean Manufacturing and the Toyota Production System (TPS). Takt time refers to the rate at which products should be produced in order to meet customer demand. If individual operations or processes are operating at a faster rate than the required takt time, the work in process or finished goods inventory will build up; this is the major contributor to waste as defined in the TPS. If individual operations or processes are unable to meet the takt time then demand will not be met. The aim is to manage all processes so that they exactly meet the takt time. Line balancing is a technique for ensuring that each process or operation has a balanced workload and can perform within the required takt time. An unbalanced line causes waste either due to operators being overworked, and hence likely to make errors, or operators waiting for the previous task to be completed. Line balancing requires a thorough analysis of the tasks being performed, including the breakdown of task time into operator time and machine time, to allow a balanced allocation of work to operators. These two techniques are often used in tandem with mixed model manufacturing, which involves scheduling multiple products so that there is constant variety along the production line rather than producing each product in large batches.

57. Target Costing

Target costing is a technique for developing products that meet both the customer's requirements for a price and the organization's requirement for a cost. It is a paradigm shift from the days when a product was developed and a profit margin was simply attached to the organization's cost to produce it. The technique was developed as a result of the recognition that the power in the marketplace was moving to the consumer, and that organizations could no longer simply set a price to cover their costs and provide the required margin. Target costing involves understanding the cost of all processes and components that go into making the product and setting a target value for each one. Efforts are then directed toward achieving the required targets.

The questions in this section of the questionnaire covered the benefits ascribable to a number of "soft" technologies directed at improving the performance of people within the organization.

58. Teams Development

See number 19 on High Performing Teams.

59. Theory Of Constraints (TOC)

The TOC is a planning and improvement technique that utilizes the principle that the throughput of a system is determined by the slowest step or the most overutilized piece of equipment (i.e., the constraint). Understanding this principle is key to any improvement activity as any effort to improve the throughput of system that does not address the constraint and will not result in an improve throughput. (Note: Throughput as defined in the TOC is the rate of conversion of raw materials into cash; that is, sold product.)

In simple terms, there are five main stages in the application of the TOC:

1. Identify the system's constraint(s): determine what is limiting the throughput of the system;

2. Exploit the constraint: maximize the use of the constrained resource;

3. Subordinate everything else to the above decision: set all other resources to the pace of the constrained resource;

4. Elevate the constraint: make changes that reduce the constraint; and

5. If the constraint has been broken, go back to step 1.

60.   Time Studies/Work Method Studies

This is an analysis of the time spent in going through the different motions of a job or series of jobs in the evaluation of industrial performance. Such studies were first instituted in offices and factories in the U.S. in the early 20th century. They were widely adopted as a means of improving work methods by subdividing the different operations of a job into measurable elements, and they were in turn used as aids in standardization of work and in checking the efficiency of workers and equipment.

61.   Total Productive Maintenance (TPM)

TPM is generally understood to stand for Total Productive Maintenance. The words are an interpretation of a translation from the Japanese referring to: All the people in the company (Total) creating greater wealth (Productive) through the caring for equipment to maximize its performance (Maintenance). Other translations have been Total Productive Manufacturing and Total Process Manufacturing. The concept is centered around a number of pillars (initially five). The original pillars were:

1. Improving equipment effectiveness (eliminating the six big losses)

2. Autonomous maintenance by operators

3. Planned maintenance

4. Training to improve operator and maintenance skills

5. Early equipment management

62. Toyota Production System (TPS)

This is the manufacturing system developed at the Toyota Motor Corporation as a means of competing against the mass production giants of the American car industry. The system is designed to minimize all forms of waste, so that the value added to a product as it moves through the manufacturing process is maximized. TPS uses a variety of tools and techniques such as Just in Time (JIT), Single Minute Exchange of Die (SMED), Kanban, leveled production, standard work, and mixed model manufacturing, to ensure that the process is responsive to the customer's needs and that inventory, poor quality, and cost are minimized.

63. Value Analysis

A value analysis involves identifying all the tasks currently conducted to complete a process, and then classifying them either as value adding, non-value adding, or business non–value adding (i.e., tasks that are necessary for operation of the business but add no value). All value is determined from the customer's perspective, so only those tasks that provide something that the customer is willing to pay for can be classified as adding value. The next stage is to redesign the process to eliminate all non–value–adding tasks, and reduce the effort required performing business non–value–adding tasks. The aim is to maximize the ratio of value–adding to non–value-adding tasks. The end result should be a more streamlined process with a much shorter cycle time.

64. Value Engineering

Value engineering involves the redesigning of products to yield the same level of performance at a lower cost. This can include redesigning to reduce the number or cost of components or to reduce the logistical or manufacturing costs. Many of the techniques used in design for manufacturing and assembly can be applied here.

65. Value Stream Mapping

Value stream mapping is a technique used to map how value is added as material moves through the production process. The total value added time can then be compared to the total time it takes for a product to flow through the production process to gauge the efficiency the system. It is a useful first step for determining the current state of operations, and also for identifying opportunities for improvement.

66.   Visual Manufacturing

Visual manufacturing refers to a system of management in which most of the necessary information to manage a process is presented in an easily assimilated visual manner. It involves techniques such as:

- Having clear visual operating instructions placed near each machine
- Setting up charts and graphs around the workplace to show the performance of individual work stations
- Developing visual standards (e.g., marks along an assembly conveyor line) that allow easy detection of when a process is operating at the correct rate
- Having a system of lights, that can be seen from anywhere in the factory, to display machine breakdowns, lack of raw materials, and so on (called Andon boards)

67.   Waste Reduction

These are techniques used to systematically identify and eliminate waste from processes. Seven major sources of waste have been categorized within the TPS, and provide the basis for uncovering and developing strategies for removing waste. The seven sources of waste are:

1. Waste of overproduction: the waste associated with making more product than is required;
2. Waste of waiting: the waste associated with waiting for material to arrive or watching a mechanized process;
3. Transportation waste: the waste associated with moving products more than necessary;

4. Processing waste: the waste associated with overprocessing of parts;

5. Inventory waste: the waste associated with producing and storing inventory;

6. Waste of motion: the waste associated with operators moving, stretching, or bending to complete tasks; and

7. Waste from product defects: the waste associated with poor quality.

68.   The Manufacturing Game®

The Manufacturing Game is designed to be a catalyst for cultural change by engaging frontline workers in bottom-up defect elimination activities. After initial training using a game play methodology that fosters teamwork and defect elimination, cross-functional action teams of between two to four people choose small defects they want to eliminate and create action plans for that purpose. A key aspect of this approach is that the teams are "action teams" not "recommendation teams." They have total ownership for implementing their plan and eliminating the defects. Considering the typical manufacturing plant has some 20,000 or more defects, eliminating them requires hundreds of small teams engaged in eliminating them. The ultimate result is an empowered workforce supporting manufacturing excellence, a major cultural shift in most organizations.

The game facilitates "little innovation": the introduction of a new idea, method, or device that will improve your processes and business. Defects that result in failures are removed or better managed. Non–value-adding activity (that costs money) is removed or minimized. Systems are optimized throughout the production or supply chain. As "little innovation" prospers, it in turn finances the "big innovation," so your business is more innovative throughout, from the CEO to the shop floor, and more successful.

69.   Kepner Tregoe®

This is a tool for problem-solving and decision-making, and helping organizations develop a rational approach to these two processes. Additional detail can be found in *The New Rational Manager* by Charles H. Kepner and Benjamin B. Tregoe, Kepner-Tregoe, Inc., Princeton, NJ, 1981.

70. 5S

This is discussed in the chapter covering Kaizen.

71. 5M

This is discussed in the chapter covering Kaizen.

72. 5 Whys

This is discussed in the chapter covering Root Cause Analysis.

73. Risk-Based Maintenance

This is a process for considering the risk of doing or not doing certain maintenance activities and using principles similar to RCM/FMEA. Risks include safety, cost, environmental, and overall business risk.

74. Proactive Maintenance

This is a process for getting to the root cause of those defects and causes that result in an unexpected maintenance requirement, help extend equipment life, or avoid failures altogether by eliminating the root cause. Note that the root cause could require a design, procurement, or operational change, and is not simply focused on maintenance.

75. Mission-Directed Work Teams

This is similar to number 18 above but has the added element of giving team members a sense of mission aligned to the corporate strategy and goals. Additional information may be obtained from David Burns, CDIA, Melbourne, Australia.

Reference

1. SIRF-Roundtables, Melbourne, Victoria, Australia.

Index